INSTRUCTOR'S MANUAL

SMITH

ECOLOGY AND FIELD BIOLOGY

FIFTH EDITION

Robert K. NEELY
EASTERN MICHIGAN UNIVERSITY

HarperCollins*CollegePublishers*

ISBN 0-06-500979-7

96 97 98 99 00 9 8 7 6 5 4 3 2 1

Table of Contents

PREFACE

How does one go about teaching a course in ecology? To a professional ecologist, this question probably seems ridiculous. Nevertheless, the variation in pedagogical approaches by seasoned ecologists is immense, usually dependent on an individual's experience and expertise. Couple this variation with courses taught by instructors with only modest training in ecology at small universities and colleges, and it becomes obvious that there is really no standard approach for instruction in the discipline of ecology. Many instructors, however, often structure their course around the levels of organization in ecology, i.e., individual, population, community and ecosystem. Within this organizational scheme, however, much leeway exists for shaping the overall context of a course. For example, a course might emphasize the functioning of systems, or population-level phenomena, or the ecological aspects of behavior, or environmental ecology, etc. The organization and focus of a course is the first decision that an instructor must make. Robert Smith's *Ecology and Field Biology* provides the flexibility for a variety of approaches.

A second question that must be answered is how to use the textbook; hopefully, this is where the instructor's manual will be of value. Instructors must first realize that there is no possibility of having students utilize the entire textbook in one semester. Rather, information relevant to classroom lectures and discussions must be chosen from the text. The overall organization of the textbook is based on six sections:

> Part I - **Introduction**: two chapters dealing with the nature of ecology in both a historical and experimental context;

> Part II - **The Organism and it's Environment**: seven chapters exploring the biological significance of various abiotic factors;

> Part III - **The Ecosystem**: three chapters emphasizing the nature of energy flow and nutrient processing in natural systems;

> Part IV - **Comparative Ecology**: a four-chapter survey of major ecosystems;

> Part V - **Population Ecology**: ten chapters devoted to important population processes;

> Part VI - **The Community**: three chapters summarizing community structure, disturbance and development.

Each chapter in this instructor's guide parallels the textbook. Instructors will first find an outline to the relevant textbook chapter with the text's page numbers for each major section; this is meant to facilitate preparation of the course syllabus and reading assignments. The outline is then followed by an overview to the content and organization of the textbook chapter. Although my commentary is based on textbook content, alternative scenarios, sources of information and organizational schemes are often offered. Furthermore, I try to provide some advice about content that I believe should be emphasized in a beginning ecology course. I also try to cross-reference information in one chapter to relevant concepts presented in other chapters of the textbook and instructor's guide.

A short list of discussion questions/topics is also provided. Although there are exceptions, in most instances these questions have an environmental theme associated with them. I've done this for two reasons. First, these questions should be used in conjunction with the review questions listed at the end of each chapter in the textbook, most of which deal with the theoretical underpinnings of an ecological topic. Second, while I don't believe it should be the focus of a beginning ecology course, I am an advocate of students learning to question the impact of their lives and human activity on the functioning of organisms and other natural systems. I also believe that students should see the relevance of ecological principles to many human actions and behaviors. Some experienced ecologists may disagree with these points, but I doubt that many will. If an ecology course doesn't stimulate a student to think about the significance of human activity in the natural scheme, then what course will? Too few students ever take courses that provide environmental awareness, e.g., environmental biology or conservation.

I conclude each chapter of the instructor's guide with a suggested lecture outline and additional resource materials (books, papers, journals, slides, films, and/or software). The outlines may or may not follow the textbook's organization. I often make recommendations about unifying information from various portions of the textbook and/or chapter, but the outlines are generally intended only as food for thought about lecture organization. The additional resource materials provide supplemental information to the textbook's content. Usually, I have tried not to list references cited in the text; however, I often emphasize particularly useful citations that an instructor may wish to consult. The titles to slides, films, etc., can enhance some subject matter, particularly if students have no familiarity with the organisms or ecological settings and they can't gain that familiarity on a firsthand basis in the course, but my inclusion of selected titles is not to be misconstrued as an endorsement of these materials. In only a few instances have I personally evaluated these items. Furthermore, ecology should be taught with an "active" mode of learning involving field and lab work, which emphasizes the process of science rather than a more passive approach based on audio-visual aids.

The instructor's guide has been prepared primarily for the occasional teacher of ecology needing some assistance in the preparation and delivery of a course, but I sincerely hope that experienced ecologists will also peruse the guide for information that may enhance their course(s). The ultimate goal for all teachers is not only to teach course content, but to also elevate the intellectual awareness and capability of students. It is my wish that both the textbook and instructor's guide will be valuable assets in this endeavor.

CHAPTER 1

ECOLOGY: ITS MEANING AND SCOPE

Text Contents	Pages
Ecology defined	*4*
Development of Ecology	*4-8*
Tensions within Ecology	*8-12*

CHAPTER SUMMARY AND ORGANIZATION

This introductory chapter provides an overview of the diversity of definitions, origins, and applications of ecology as a scientific discipline. The chapter is divided into three major sections which describe in order (1) definitions of ecology, (2) development of ecology, and (3) disputes among ecologists. In the first section, the origin and definition of ecology as a scientific discipline are explored. Ten contemporary definitions of ecology are presented in addition to Ernst Haeckel's original definition. Ecology as "the study of the structure and function of nature" is suggested as an adequate working definition.

The development of ecology into a recognizable discipline is detailed in the second section of Chapter 1. Clearly, ecology has evolved as a science because of the contributions of many individuals from an array of backgrounds. As a result of this diverse heritage, ecology evolved into a variety of specializations, such as theoretical mathematical ecology, behavioral ecology, evolutionary ecology, physiological ecology, chemical ecology and others. The important historical facets of plant ecology, animal ecology, physiological ecology, population ecology, applied ecology and cooperative efforts among individuals from diverse ecological subdisciplines are described separately in this section. The early contributions of plant geographers, aquatic biologists, zoologists, animal behaviorists, and natural historians, as well as the significance of Charles Darwin's theory of evolution and Gregor Mendel's work on inheritance, are detailed.

The third section of Chapter 1 discusses some of the controversies and divisions that have arisen during the development of ecology. Most notable in this discussion are the events that led to long-term separation between plant ecology and animal ecology, organismal versus individualistic theories of plant communities, and between ecosystem ecology (a holistic approach to ecology) and population ecology (a reductionist approach to ecology). This section ends with a contrast of theoretical ecology and applied ecology; the latter being the application of ecology to solving human-induced environmental problems. This section has a strong environmental theme and a historical perspective is presented about the development and diversification of applied ecology into the fields of wildlife management, conservation biology, landscape ecology and restoration ecology.

Several themes can be drawn across the four sections of Chapter 1 that should be emphasized in the classroom. First, ecology is a multifaceted discipline with a complex history. The breadth of topics across which ecology ranges is probably unparalleled among scientific disciplines. Ecology, particularly a course in general ecology, will range across biology, chemistry, physics, hydrology, mathematics, and many other disciplines. Biological topics alone cover aspects of evolution, physiology, reproduction, behavior, genetics and anatomy for both plants and animals. While this diversity has naturally resulted in differences of opinion among ecologists, it also has resulted in the accumulation of an immense amount of information relevant not only to the development of ecological theory but also to the solution of the overwhelming environmental problems plaguing our planet. Despite the multifaceted nature of ecology, it is rightfully a biological discipline. We are interested primarily in living organisms in terms of how they respond to and affect their environment.

Secondly, ecology should be emphasized as a science of interactions. In basic terms, organismal behavior, abundance, distribution, evolution, etc., are closely intertwined with a variety of environmental resources and constraints, both living and non-living. Finally, the existence, nature, and magnitude of historical and recent disputes in ecology are not topics to be ignored in the classroom. Students need to learn that multiple points of view are typical in any discipline and that these differences often foster intellectual development and progress. The presentation and discussion (and maybe even debate) of alternative viewpoints in ecology can be a useful teaching tool by which students learn to think critically.

TOPICS FOR DISCUSSION

1. How varied are the kinds of questions that ecologists might investigate? What are some contexts in which ecological questions can be formulated? What are examples of ecological questions?

2. How do evolutionary processes relate to the study of ecological processes?

3. What are the major environmental concerns facing humanity? Do clear-cut answers to Earth's environmental problems exist; if not, what contributes to the complexity of how to handle these problems? What can (or has) ecology contributed to environmentalism and what should the role of ecologists be in environmental protection and restoration?

4. What are the positive and negative aspects of disagreement and debate among ecologists about ecological phenomena?

5. Should ecology be characterized as a precise science? If not, what factors constrain the ability of ecologists to always make definitive predictions or statements?

LECTURE OUTLINE

I. Ecology: Origin and Definition

A. Historical Overview
    1. Ernst Haeckel's Definition of "Okölogie"
    2. Contributions of Early Plant Geographers
    3. Significance of Darwin's and Mendel's Work
    4. Evolution of Energetic Studies
    5. Contributions of Animal Biologists
B. Contemporary Ecology
    1 Definition(s)
    2. Organizational Levels
        a. Biological
            i.   individual
            ii. population
            iii. community
            iv. ecosystem
        b. Ecological
            i.   evolutionary ecology
            ii. behavioral ecology
            iii. physiological ecology
            iv. systems ecology
            v.   theoretical ecology
            vi. applied ecology

LECTURE OUTLINE CONTINUED:

II. Classic Disputes in Ecology

      A. Early Division Between Plant and Animal Ecologists
      B. Organismal Versus Individualistic Plant Community Concepts
      C. System-based Approach and Population-based Approach to Ecology

RESOURCE MATERIALS

Reference Material

The citations included in Chapter 1 reference many of the original and significant papers that have shaped the field of ecology. In addition to these references, the following books and papers may provide additional insight into the intellectual lineage of this discipline. A compilation of historically significant ecological articles can be found in Real and Brown (1991) while the remaining references provide a historical perspective about the development of contemporary ecology. McIntosh (1985), Worster (1977) and Hagen (1992) are particularly useful.

Egerton, F.N. (editor). 1977. *History of American Ecology*. Arno Press, New York.

Golley, F.B. 1993. *A History of the Ecosystem Concept in Ecology*. Yale University Press, New Haven, CT.

Grove, R.H. 1992. "Origins of Western Environmentalism." *Scientific American* 267(1):42-47.

Hagen, J. 1992. *An Entangled Bank - The origins of ecosystem ecology*. Rutgers University Press, New Brunswick, NJ.

Kingsland, S. 1985. *Modeling Nature*. University of Chicago Press, Chicago, IL.

McIntosh, R. 1985. *The Background of Ecology*. Cambridge University Press, New York, NY.

Real, L.A. and J.H. Brown (editors). 1991. *Foundations of Ecology: Classic Papers with Commentaries*. The University of Chicago Press, Chicago, IL.

Worster, D. 1977. *Nature's Economy*. Sierra Club Books, San Francisco, CA. An engaging history of ecology, a different view of history from that of McIntosh.

Multimedia Resources

Insight Media, 121 West 85th Street, New York, NY, 10024. --- **"What is Ecology"** - this video provides a generalized definition of an ecosystem and examines interrelationships between plants and animals, flows of energy, biomes, and applied ecology.

JLM Visuals, 1208 Bridge Street, Grafton, WI 53204. --- **"Introduction to Ecology"** - this set of 35mm. slides provides an overview to fundamental ecological concepts.

Rex Educational Resources Company, 2700 York Road, Burlington, NC 27215. --- **"What Ecology Is and Is Not Set"** - this set of 35mm. slides provides insight into misunderstandings about ecology and the kinds of questions that ecologists ask.

CHAPTER 2

ECOLOGICAL EXPERIMENTATION AND MODELS

CHAPTER SUMMARY AND ORGANIZATION

First, and foremost, ecology is a science; hence, the essence of Chapter 2 is to relate the essential components of scientific reasoning, i.e., the "scientific method," to the discipline of ecology. At the simplest level, science is separated from other intellectual pursuits by two factors: (a) the content or knowledge base related to a particular scientific discipline and, perhaps more importantly, (b) *the procedure of science*. This second introductory chapter details the *procedure* of ecology as a scientific discipline. After a brief introduction to inductive and deductive logic, this chapter relates the primary aspects of the scientific method to the exploration and understanding of ecological phenomena: the collection of data for hypothesis testing, experimentation, inductive and deductive reasoning, and modeling.

The second and third sections of Chapter 2 provide insight into data collection (observations and experimentation) and hypothesis testing, respectively. Depending on one's point of view, teachers may want to consider these sections as an integrated unit related to data collection and data analysis within the context of hypothesis testing. Data can be collected by direct observation of natural events or systems, by observation of laboratory experiments, and/or by observation of field experiments. Generally, the laboratory and field experiments are designed to manipulate some important variable(s) while other variables are kept constant. The advantages and disadvantages to the various data collection approaches are also described. The section regarding hypothesis testing provides an introduction to basic statistical theory and the statistical risks associated with the acceptance/rejection of hypotheses. Together, these two sections contain many important considerations; in particular, differences between dependent versus independent variables, control versus experimental treatments, independent replication versus pseudoreplication and null versus alternative hypotheses should be emphasized.

The final segment of Chapter 2 is a detailed overview of ecological models and their value. A model is defined as "an abstract representation of the real system for the purposes of predicting the response of a dependent variable(s) based on an explicit assumption or set of assumptions." The section is organized around the presentation of statistical models, analytical models and simulation models (the latter two of which represent nonstatistical models). The logistic growth model, in its differential form, is used as a simple example of an analytical model, which can be solved mathematically and which predicts the response of a group of organisms. In contrast, simulation models do not lend themselves to analytical solutions and can be formulated in a number of ways; the text details one class of simulation models, individual-based population models, which simulate each organism in a population rather than the population as a whole. A forest gap model for prediction of the important events in the life history of forest trees is used as an example of a simulation model. Ultimately, models of any type are of the most value if they can be validated with data from a real-life system. Even unvalidated models might still prove valuable by providing new interpretations or insights to a problem. In summary, this section is particularly relevant to contemporary ecology and will provide a basis for understanding the models of population growth, competition, and predation, which are discussed in later chapters.

In any science class, including one in ecology, the fundamentals of the scientific procedure need to be continually reinforced. For a student of ecology, a fundamental understanding of the information presented in Chapter 2 provides a basis for (1) understanding example studies and various models presented later in the book, (2) further reading of the ecological literature, (3) converting ideas or questions into testable hypotheses, and (4) designing experiments to particular ecological questions. In fact, for an ecology course with a laboratory component, this chapter is especially relevant. One particularly meaningful laboratory approach is to use this chapter as a foundation for allowing students to test simple hypotheses by collecting data according to an appropriate experimental design. Ideally, these hypotheses and/or data should be integrated with information from the literature to arrive at a clear understanding and interpretation of the ecological question being considered.

TOPICS FOR DISCUSSION

1. What is the scientific method? How much variation seems to exist in the experimental approaches used in contemporary ecological studies?

2. Provide several groups of students with an ecological question and the same background information. Ask these groups to formulate hypotheses to test the question and to devise appropriate experimental designs to address these hypotheses. How similar or dissimilar are the hypotheses and experimental approaches between these student groups? What types of reasoning are used to formulate the experimental protocol? Can these students discuss or develop a verbal model that might be useful to their experimental approach?

3. Do models need to take into account the degree of environmental variability that may exist in a natural system? Why or why not? What kinds of variation can be easily identified?

4. To what extent will environmental heterogeneity, temporal or spatial, affect the design of experiments and analysis of data?

LECTURE OUTLINE

I. The Procedure of Science as Applied to Ecology

    A. Theories
    B. Hypotheses
    C. Experimental Design
        1. Types of Reasoning
            a. Inductive logic
            b. Deductive logic
        2. Dependent and Independent Variables
        3. Hypothesis Formulation and Testing
            a. Null vs. Alternative Hypotheses
            b. Type I vs. Type II errors

II. Ecological Modeling

    A. Value of Models
        1. Prediction
        2. Conceptualization
        3. Hypothesis Formulation

LECTURE OUTLINE CONTINUED:

      B. Model Types
          1. Conceptual and Graphic Models
          2. Statistical Models
          3. Nonstatistical Models
              a. Analytical Models
              b. Simulation Models
     C. Validification and Verification

RESOURCE MATERIALS

Reference Material

In addition to the references provided in Chapter 2, the following books and papers provide reasonable overviews to the essential elements of biological and ecological experimental design. In particular, the classic paper by Platt (1964) reinforces the idea that productive scientific efforts are based on inductive inference and systematic methods. The application of scientific reasoning or "inquiry" in the classroom and the essentials of teaching students how to effectively critique research are presented in Uno (1990) and Kuyper (1991). Additionally, the reading from the National Academy of Science is an excellent introduction to the role of research and the methods of scientists, as well as other relevant information (Committee on Conduct of Science, 1989).

In their entirety, some of these references provide a much greater statistical orientation than would be appropriate for an introductory ecology course, e.g., Green (1979), Ludwig and Reynolds (1988), Mead (1988) and Turner and Gardner (1991). These books, however, represent excellent reference materials for the design of field and laboratory experiments and the analysis of data sets. At the least, Fowler (1990) is a good overview of statistical pitfalls to avoid in conducting ecological experiments.

The application of models to ecology has been extensive and models of one type or another have been generated for population dynamics, various aspects of lake and wetland ecosystems, animal behavioral patterns, microbial processes, and nutrient movements, just to name a few. Consequently, the literature about ecological modeling is vast and often difficult to read. Nevertheless, Kitching (1983) and Hall and Day (1977) are useful because they provide a general overview to modeling and its uses. For detailed presentations of models that should be discussed in an introductory ecology course, Rose (1987) is good. In addition to a number of recent monographs (e.g., DeAngelis and Gross, 1992), one journal published by Elsevier Science Publishers, *Ecological Modelling*, is a source for specific examples of models and additional literature.

Committee on the Conduct of Science. 1989. *On Being a Scientist*. National Academy Press.

DeAngelis, D.L. and L. J. Gross (editors). 1992. *Individual-based Models and Approaches in Ecology : Populations, Communities, and Ecosystems*. Chapman & Hall, New York.

Fowler, N. 1990. "The 10 most common statistical errors." *The Bulletin of the Ecological Society of America* 71:161-164.

Green, R.H. 1979. *Sampling Design and Statistical Methods for Environmental Biologists*. John Wiley and Sons, Inc., New York.

Hall, C.A.S. and J.W. Day. 1977. *Ecosystem Modeling in Theory and Practice: An Introduction with Case Histories*. John Wiley and Sons, Inc., New York.

Kitching, R.L. 1983. *Systems Ecology - An Introduction to Ecological Modelling*. University of Queensland Press, New York.

Krebs, C.J. 1988. *Ecological Methodology*. Harper and Row Publishers, New York.

Kuyper, B.J. 1991. "Bringing up scientists in the art of critiquing research." *Bioscience* 41:248-250.

Ludwig, J.A. and J.F. Reynolds. 1988. *Statistical Ecology*. John Wiley and Sons, Inc., New York.

Mead, R. 1988. *The Design of Experiments: Statistical Principles for Practical Applications*. Cambridge University Press, New York.

Platt, J.R. 1964. "Strong inference." *Science* 146:347-353.

Rose, R.R. 1987. *Quantitative Ecological Theory: An Introduction to Basic Models*. John Hopkins University Press, Baltimore.

Turner, M.G. and R.H. Gardner. 1991. *Quantitative Methods in Landscape Ecology*. Springer-Verlag, Inc., New York.

Uno, G.E. 1990. " Inquiry in the classroom." *Bioscience* 40:841-843.

Multimedia Resources

Educational Images, P.O. Box 3456, West Side, Elmira, NY 14905. --- **"Scientific Method"** - a slide set detailing the use of experimental study in scientific disciplines.

CHAPTER 3

ADAPTATION

CHAPTER SUMMARY AND ORGANIZATION

From its inception, ecology has been an evolutionary science. Chapter 3 provides the initial foundation for understanding the structure, distribution and functioning of organisms by placing a few basic aspects of evolution and adaptation into an ecological context. In the three short sections of this chapter, adaptation is described and related to the upper and lower environmental limits constraining organisms and to their need for homeostatic maintenance of their internal environment.

The first topic on adaptation provides the basis for understanding the remaining topics and their relevance to ecology (particularly Chapter 21). Three general definitions of adaptation are initially presented; the latter of the three is perhaps the most useful for an introductory ecological course in which organismal/species-level response to environmental pressures is a course theme. Adaptations have typically arisen because of and result in long-term, elevated reproductive success for individuals carrying the adapted traits. As the textbook emphasizes, students should understand that the important criterion for reproductive success is the number of descendants that are left by an organism rather than the number of offspring produced. This idea must be emphasized to beginning students, for it is only in this context that the components of fitness and natural selection can be understood.

In some cases, an "adaptation" may not have arisen as a specific evolutionary response to the specific set of environmental conditions in which it seems useful. Clearly, these events can complicate interpretations of adaptation. Despite this problem, the ideas that must be emphasized to students are: (1) adaptability depends on genetic variability; (2) genetic variability is essential for selection processes to occur. Students should become familiar with the ideas of genotypic, phenotypic and ecotypic variation and that the lack of genetic variability restricts the ability of populations to respond to environmental change through natural selection.

The second section of this chapter on adaptation begins by noting that adaptability can also be considered in terms of the range of environmental conditions in which an organism can function, in other words, an organism's "tolerance limits." In this context, three important laws relating to maximum and minimum environmental limits on organisms are presented. First, Liebig's Law of the Minimum is presented, in which the growth of an organism is proportional to the single most limiting environmental requirement. Students must be cautioned that this law only applies under equilibrium conditions. If all other environmental conditions are held constant, an organism will exhibit a positive response to increasing availability of the most limiting environmental resource until that resource is no longer limiting and some other environmental resource has become limiting.

Second, the Law of Limiting Factors is an extension of the Law of the Minimum that includes a recognition that organismal functioning may also be affected, or limited, by maximum quantities of a resource. Third, this law is extended into a consideration of the Law of Tolerance, in which organisms are constrained by both the maximum and minimum extremes of an environmental condition; thus, these extremes represent the limits of tolerance. This idea is described in detail by the use of two figures that illustrate tolerance limits. Furthermore, the terminology associated with tolerance limits is briefly explained with regard to wide tolerances (eury-) and narrow tolerances (steno-). Two points should be accentuated: (1) In general, organisms with wide ranges of tolerance to many environmental factors will have a wider distribution than organisms with narrow ranges of tolerance; (2) the range of tolerance is a function of the adaptations that an organism possesses for coping with specific environmental components; hence, organismal distribution is clearly determined by specific adaptations.

Information about homeostasis in organisms concludes Chapter 3. It is essential for students to understand that, given the overall magnitude of environmental variation, many organisms are capable of regulating their internal environments to some extent by a combination of physiological, morphological and behavioral means. Temperature regulation is used in the text as an example of the extent to which organisms may or may not regulate specific physiological parameters (see Chapter 6 for a more detailed presentation of temperature regulation). Consequently, homeotherms are contrasted with poikilotherms. Essential concepts to present in the classroom relate to the maintenance of homeostasis by negative and positive feedback mechanisms.

In summary, this chapter weaves a thread of continuity among ideas about adaptation, tolerance, and homeostasis. Conceptually, students should come away from this chapter with the understanding that: (1) adaptations usually arise through natural selection as a way of coping with environmental rigor and that these adaptations confer fitness to organisms; (2) the range of conditions that an organism can tolerate depends on the adaptations possessed by an organism for dealing with those conditions; (3) given that all organisms maintain some degree of homeostasis, tolerance ranges represent the range of conditions under which homeostatic mechanisms can function; (4) homeostasis requires specific physiological, behavioral and morphological capabilities, which are themselves adaptations for dealing with the environment. In later chapters, the ecological significance of homeostasis becomes evident, i.e., Chapter 5 (water balance), Chapter 6 (temperature regulation), and Chapters 18 and 19 (population regulation).

TOPICS FOR DISCUSSION

1. To paraphrase a statement by Paul Colinvaux in his *Ecology* textbook, a species is not a group of individuals classified as possessing a common shape, but a population of individuals suited for a common way of life to which their common shape is adapted. What does this statement mean, and to what extent does it have merit? How do the ideas of adaptation, fitness and natural selection apply here?

2. It has been said that "adaptations reflect the past while natural selection occurs in the present." What does this mean?

3. What is wrong with the assumption that species are continually improving through the acquisition of adaptations?

4. How does the law of tolerance relate to organismal response to pollution or other human-induced impacts on ecosystems?

5. In an evolutionary context, what are the advantages of wide ranges of tolerance and narrow ranges of tolerance for environmental factors? What kinds of environments may select for wide tolerance ranges as opposed to narrow tolerance ranges?

LECTURE OUTLINE

I. Natural Selection and Adaptation

    A. Components of Natural Selection
        1. Genetic Variation
        2. Inheritance
        3. Environmental Selection
    B. Adaptation and Fitness

II. Environmental Limits on Organisms

    A. Liebig's Law of the Minimum
    B. Law of Limiting Factors
    C. Shelford's Law of Tolerance
        1. Prefixes
            a. eury-
            b. steno-
        2. Tolerance and Distribution

III. Homeostasis

    A. Maintenance of the Internal Environment
    B. Homeostasis and Feedback Mechanisms
        1. Positive Feedback
        2. Negative Feedback

IV. Discussion of the Relationships Between Adaptation, Tolerance, and Homeostasis

RESOURCE MATERIALS

Reference Material

For an additional introduction to the subjects of adaptation, tolerance and homeostasis, Chapters 1 and 6 in Pianka (1988) provide examples and insights as well as the citations to Liebig's and Shelford's original works. Similarly, Krebs (1985) considers these ideas in the overall context of how the distribution of organisms is determined. Lewontin (1978) is also an excellent introduction to the idea of adaptation. For very detailed readings with extensive bibliographies, Endler and McLellan (1988) and McDonald (1983) give a thorough analysis of adaptation with an emphasis on the coupling of molecular and genetic components to overall evolutionary processes. The remaining references are generally useful for examples and other information relevant to Chapter 3.

Endler, J.A. and T. McLellan. 1988. "The processes of evolution: Toward a newer synthesis." *Annual Review of Ecology and Systematics* 19:395-421.

Errington, P.L. 1956. "Factors limiting higher vertebrate populations." *Science* 124:304-307.

Lewontin, R.C. 1978. "Adaptation." *Scientific American* 239:213-230.

Krebs, C.J. 1985. *Ecology - The Experimental Analysis of Distribution and Abundance.* Harper and Row, Publishers, New York.

McDonald, J.F. 1983. "The molecular basis of adaptation: A critical review of relevant ideas and observations." *Annual Review of Ecology and Systematics* 14:77-102.

Pianka, E.R. 1988. *Evolutionary Ecology* (4th edition). Harper and Row, Publishers, New York.

Ruibal, R. and R. Philibosian. 1970. "Eurythermy and niche expansion in lizards." *Copeia* 1970:645-653.

Schmidt-Nielsen, K. 1975. *Animal Physiology: Adaptation and Environment*. Cambridge University Press, London.

Multimedia Resources

JLM Visuals, 1208 Bridge Street, Grafton, WI 53204. --- **Plant and Animal Adaptations -** twenty slides presenting various behavioral and physical adaptations.

CHAPTER 4

CLIMATE

CHAPTER SUMMARY AND ORGANIZATION

An understanding of the primary determinants of climate is essential to understanding the behavior, physiology and distribution of plants and animals. Chapter 4 provides a basis for later chapters in the text (e.g., chapters on water balance, temperature, and biome descriptions) by describing the primary determinants of global and regional climate. Climate (combined temperature, moisture, precipitation and wind regimes) should be considered the product of weather patterns averaged over time. In this chapter, climate is discussed at increasingly smaller spatial scales; global climatic patterns are initially considered, followed by a consideration of regional climates and finally a summary of the significance of the microclimate concept. The chapter ends with a discussion of plant distribution and climatic patterns.

As an alternative organization, parts of Chapter 4 could be postponed for later consideration in the classroom. For example, because microclimate is essentially small-scale differences in moisture, temperature and light, a discussion and/or series of activities could be organized to utilize information presented in Chapters 5 (Water Balance), 6 (Heat Budget) and 7 (Light). Furthermore, the consideration of life-zones and biomes could be deferred and integrated into later material regarding ecosystem types and characteristics (Chapters 13-16).

The foundation for understanding global climate is presented in the first two sections of Chapter 4. In essence, Earth functions as a huge heat engine powered by solar energy. Heat energy is not distributed equally over the planet; hence, variations in heat budgets ultimately account for global temperature, wind and rain patterns, as well as ocean currents. Students should understand how adiabatic processes associated with rising and sinking air masses combine with the coriolis force to account for major planetary patterns of wind and precipitation/evaporation. Additionally, the winds associated with a moving atmosphere and the coriolis force are the driving forces behind major oceanic currents. Ocean currents, in turn, significantly alter global climate, as illustrated by the discussion of La Nina and El Nino events.

The section describing regional climates is subdivided into sections regarding the effects of topography and descriptions of humidity and inversions, the latter of which are obviously coupled with topographic influences. Students should come away from this section with an understanding that the climate of a particular region is not only a function of global climatic patterns, but is also influenced by a multitude of other factors, e.g., position in a continental land mass, topography, absence or presence of nearby bodies of water, etc. In particular, the topography of mountainous regions can have pronounced effects and may result in rainshadow effects, inversion patterns, valley fog, and upslope and downslope winds. Large bodies of water, moderate climatic regimes and coastal environments may result in marine inversions.

At the smallest scale, microclimates represent "small" climates that differ from the overall prevailing climatic conditions because of differences in topography, height above the ground, plant cover, aspect, etc. The discussion of microclimates is organized around five sections describing sequentially north- versus south-facing slopes, the microclimate near the ground, the influence of vegetation on microclimate, microclimate as an important habitat variable, and urban microclimatic settings. For example, north-facing slopes and south-facings slopes may be very different with respect to temperature, moisture and humidity patterns. Students should understand that microclimates represent important habitats for plants and animals. As humans, we tend to overlook the importance of microclimate because we are masters at creating optimal microclimates for ourselves inside our homes, workplaces, and even cars through the use of thermostats, heaters, and air conditioners. Animals seek out favorable conditions and much of their behavior is affected by the presence and absence of favorable microclimatic conditions. Plants, too, are influenced by microclimatic conditions and depend on favorable microclimatic sites for colonization.

The final section of Chapter 4 examines the distribution of terrestrial plant communities over the Earth. First, a general description of major vegetative zones is described. Primarily, the location of these zones is a function of two climatic components, precipitation and temperature. This section concludes with various ways to consider the classification of large vegetation units. Biomes, the Holdridge life zone system, and ecoregions are sequentially discussed. The biome concept may be most useful to students in an introductory ecology course.

TOPICS FOR DISCUSSION

1. Why do many deserts occur along west coasts with cold currents, e.g., South America and Africa? What is the role of inversion patterns in the formation of these deserts?

2. The Galapagos Islands are on Earth's equator in the Pacific Ocean, about 600 miles off of the coast of Ecuador. These islands are arid and dominated by xerophytic vegetation. Given the latitudinal position of these islands, why aren't they covered in lush, tropical vegetation? What are the primary determinants of local climate on the Galapagos Islands?

3. What topographic, oceanic, or latitudinal factors account for the location of many of Earth's deserts?

4. Microclimate has been described as that climate in contact with an organism's surface. What does this mean? Would you agree with or disagree with this statement? Why?

5. Why are plants good indicators of climatic conditions?

6. Would you expect mountainous regions to have a wider array of microclimatic conditions than a flat, grassland region? Why? How would you expect microclimate to vary in each habitat type? What would be the primary determinants of microclimate?

7. What are some animal behaviors that seem closely coupled to use of microclimates? Have you observed any local plant distributions that might suggest changing microclimatic conditions?

LECTURE OUTLINE

I. Solar Radiation

    A. Atmospheric Reflection and Absorption
    B. Albedo
    C. Latitudinal Variation in Heat Distribution

LECTURE OUTLINE CONTINUED:

II. Global Atmospheric and Oceanic Currents

A. Wind Patterns
  1. The Adiabatic Process
  2. Coriolis Force
  3. Global Patterns
    a. Tradewinds
    b. Prevailing Westerlies
    c. Polar Easterlies
B. Ocean Currents
  1. Role of Atmospheric Winds
  2. Coriolis Force and Gyres
  3. Effects on Heat Budgets
    a. Redistribution of Heat by Currents
    b. Areas of Cold, Oceanic Upwelling
    c. La Nina versus El Nino Events

III. Regional Climates

A. Topographic Influence
  1. Rain Shadow
  2. Inversions
B. Proximity to a Water Body
  1. Continental versus Maritime Climates (not discussed in chapter)
  2. Lake Effect (not discussed in chapter)
  3. Marine Inversions
C. Humidity
  1. Vapor Pressure
  2. Vapor Pressure Deficit
  3. Absolute Humidity
  4. Relative Humidity
    a. Influence of Temperature & Wind
    b. Diurnal Cycle

IV. Microclimate

A. Examples
B. Importance to Plants and Animals

V. Climate and Vegetation

A. Altitudinal and Latitudinal Zonation
B. Biomes
C. Holdridge Life Zone System
D. Ecoregions

RESOURCE MATERIALS

Original references are cited in the text of Chapter 4 for Merriam Life zones (Merriam, 1894a; 1894b), biomes (Clements and Shelford, 1939), ecoregions (Bailey, 1976) and the Holdridge life zone system (Holdridge et al., 1971). In addition to these references, the following books contain useful information about general climatological mechanisms, microclimatic factors, and the distribution of plants and animals, particularly with regard to climate, i.e., biogeography. In particular, Akin (1990) provides an excellent overview of present and past climates with emphasis on global vegetation patterns and soil distributions (the information about soils will also be relevant to Chapter 9 regarding soils). Additionally, Rosenberg's book (1983) provides many useful perspectives to understanding aspects of microclimate; again, much of this information will be useful in later chapters, particularly the chapter regarding heat budgets (Chapter 6). Finally, it would not be inappropriate to include information about prospective global climate change and its effect on biological systems in lectures about the ecological significance of climate; thus, Volume 23 (1992) of the *Annual Review of Ecology and Systematics* includes a special section on Global Environmental Change in which nine papers detail important considerations of climatic change. Ford (1982) also provides an interesting summary of possible biological alterations in response to climate change.

Akin, W.E. 1990. *Global Patterns: Climate, Vegetation and Soils*. University of Oklahoma Press, Norman, OK.

Cox, C.B. and P.D. Moore. 1995. *Biogeography: An Ecological and Evolutionary Approach* (5th edition). Blackwell Scientific, Boston, MA.

Brown, J.H. and A.C. Gibson. 1983. *Biogeography*. The C.V. Mosby Company, St. Louis, MO.

Ford, M.J. 1982. *The Changing Climate: Responses of the Natural Flora and Fauna*. George Allen and Unwin, London.

Jones, H. G. 1992. *Plants and Microclimate: A Quantitative Approach to Environmental Plant Physiology* (2nd Edition). Cambridge University Press, New York.

Rosenberg, N.J. 1983. *Microclimate: The Biological Environment* (2nd Edition). John Wiley and Sons, Inc., New York.

CHAPTER 5

WATER BALANCE

CHAPTER SUMMARY AND ORGANIZATION

This chapter provides a basis for understanding the ways in which water influences life and various ecological phenomena. Most students are probably familiar with the importance of water to various physiological processes in plants or animals; few students, however, are probably completely aware of the pervasive influence that water has on the natural world. At the most fundamental level, the requirement for water profoundly affects the distribution, physiology, and anatomy of plants and animals, as well as the structure of terrestrial and aquatic ecosystems.

The first two sections of Chapter 5 describe many of the structural and chemical properties of water. Emphasis should be placed on this information because these properties are directly related to the underlying reasons for the many effects of water on the environment and organisms. In particular, students should be aware of the significance of hydrogen bonding, lattice structure, viscosity, surface tension, specific heat, adhesion, cohesion and the relationship between water temperature and density.

Both the local and global water cycles are described in the third section of Chapter 5. Students should understand that while we do live on a water planet (~71% of Earth's surface), only a tiny amount of this water is available to life as freshwater; hence, much of the biosphere depends on the continual cycle of renewal and replenishment of freshwater. After an initial description of the various reservoirs of water, this section details the local water cycle. The processes of precipitation, evaporation, transpiration, interception, infiltration, and percolation are explained, as well as overland flow and the partitioning of soil moisture into capillary water and hygroscopic water. An interesting summary of the effects of urbanization on some of the processes in the local water cycle is also given. The final segment of this discussion on water cycles ends with a consideration of the global water cycle with an overall geographic perspective. This perspective includes discussion of the overall patterns of evaporation and precipitation over land and ocean and the influence of global wind patterns.

Chapter 5 ends with two sections that describe a variety of ways in which the biota of habitats is (or has been) influenced by water. These sections constitute more than half of this chapter and deal sequentially with plants and then animals. In the sections regarding plants, the discussion proceeds from a description of water uptake and balance to a general consideration of the short-term responses and long-term evolutionary adaptations to both drought and flooding with numerous examples.

The final section emphasizes the water balance of animals and their responses and adaptations to drought and flooding. In comparison with plants, the mobility of animals allows them to seek more favorable habitats during periods of suboptimal moisture conditions. Nonetheless, the mechanisms of water balance are much more elaborate in animals than plants and involve both passive and active transport mechanisms across membranes in a variety of structures, e.g., specialized gills, kidneys, etc. Additionally, various animals conserve water by tolerating hyperthermia, controlling respiration, or possessing various behavioral or anatomical characteristics, e.g., estivation, salt glands, etc. The effects of moisture availability, however, are not limited to behavioral and physiological adaptations. Drought, in particular, can alter food selection in herbivores, result in outbreaks of herbivorous insects, alter mortality and fecundity, and slow insect development. Alternatively, excessive moisture spreads disease among both animals and plants by promoting the spread of fungi, bacteria and viruses.

TOPICS FOR DISCUSSION

1. Would you agree with the statement that water is the basic medium of life? Why or why not?

2. Of significance to aquatic organisms is the fact that ice floats. Do you think that aquatic organisms would be as abundant if ice did not float in cold regions?

3. Should development plans for a site, i.e., urbanization, include consideration of the potential effects on the local hydrologic cycle?

4. Given that plants need to open their stomata to facilitate both water uptake and CO_2 uptake, what are some probable effects of the global warming on water balance in plants?

5. The density of water can provide buoyancy but impede movement of organisms because of its viscosity. How have some animals adapted to either utilize or overcome these properties of water?

6. Explain how a cactus is well-adapted to arid environments. Consider the cacti's shape, spines, and growth rate. Include in this consideration not only the effects of these factors directly on water conservation, but also how these factors may affect the heat budget, which may in turn alter evaporation rates (see Chapter 6 for information on heat budgets).

LECTURE OUTLINE

I. Physical and Chemical Properties of Water

    A. Water Structure
       1. Covalent Bonds
       2. Molecular Polarity
       3. Hydrogen Bonding
       4. Lattice Structure
    B. Physical Characteristics
       1. Density
       2. Specific Heat
       3. Viscosity
       4. Surface Tension

II. Water Cycle

    A. Global Distribution of Water
    B. Important Processes
       1. Evaporation
       2. Precipitation
       3. Interception
       4. Percolation
       5. Infiltration
       6. Overland Flow
    C. Global Cycle
    D. Local Cycle

LECTURE OUTLINE CONTINUED:

III. Water Balance in Plants and Animals

    A. Plants
        1. Soil - Plant - Atmosphere Continuum
        2. Conservation of Water
            a. Temporary Responses
            b. Anatomical Adaptations
            c. Physiological Adaptations
        3. Response to Flooding
    B. Animals
        1. Saline Habitats and Salt Accumulation
        2. Arid Habitats and Water Conservation
            a. Short-term Responses
            b. Anatomical Adaptations
            c. Physiological Adaptations

RESOURCE MATERIALS

Many valuable references are cited in the text. In particular, Schultz et al. (1987) and Kramer (1983) are useful for understanding plant water relations. In addition to these references, Berner and Berner (1987) provide an in-depth overview to the properties of water, the distribution of water, and various aspects of the global water cycle. Further information about water relations in plants can be found in Simpson (1981), Meidner and Sheriff (1976), and Barbour et al. (1987). In particular, Lange et al. (1976) includes detail about water balance in plants of different photosynthetic types (C_3, C_4 and CAM). For vertebrates, Pough et al. (1989) and Schmidt-Nielsen (1990) provide an excellent overview to osmoregulation in vertebrates and invertebrates. Additionally, perusal of the journal, *Physiological Zoology,* will yield a variety of references about osmotic and ionic regulation in animals.

Barbour, M.G., J.H. Burk, and W.D. Pitts. 1987. *Terrestrial Plant Ecology*. The Benjamin/Cummings Publishing Co., Inc., Menlo Park, CA.

Berner, E.K. and R.A. Berner. 1987. *The Global Water Cycle - Geochemistry and Environment.* Prentice-Hall, Inc., Englewood Cliffs, NJ.

Lange, O.L., L. Kappen and E.D. Schulze. 1976. "Water and Plant Life: Problems and Modern Approaches." *Ecological Studies Volume 19.* Springer-Verlag, NY.

Meidner, H. and D.W. Sheriff. 1976. *Water and Plants.* John Wiley and Sons, Inc., New York.

Pough, H.F., J.B. Heiser, and W.N. McFarland. 1989. *Vertebrate Life* (3rd Edition). Macmillan Publishing Co., New York.

Schmidt-Nielsen, K. 1990. *Animal Physiology: Adaptation and Environment* (4th Edition). Cambridge University Press, New York.

CHAPTER 6

THERMAL BALANCE

CHAPTER SUMMARY AND ORGANIZATION

In addition to water balance (described in Chapter 5), organisms must maintain an optimal thermal balance. This chapter is essential to understanding many of the morphological and behavioral adaptations of animals and/or plants. Initially, an overview to heat budgets and the influence of temperature on metabolism are presented in Chapter 6 before (1) specific ways in which plants and animals can modify their heat budgets are detailed and (2) the general influence of temperature on organismal distribution is described.

The first section of this chapter establishes a context, heat budgets, in which various plant and animal responses to temperature can be evaluated. Heat budgets can be used to describe and quantify the thermal balance of an organism. Organisms, particularly ectotherms, have some control over their body temperature by being able to manipulate their heat budget. Students must understand the following concepts: (1) the ultimate source of heat is solar radiation; (2) at its fundamental level, a heat budget means that heat energy gained by an organism must equal the heat energy lost and energy stored; (3) heat gain is the total of heat inputs from the sun, surrounding environment and metabolism, while heat loss is the sum of infrared radiation (sometimes called reradiation), conduction, convection and evaporation. Assuming no heat storage, then in the most simple form, thermal balance for an organism can be expressed as follows:

Metabolic Heat + Radiant Heat = Radiation + Conduction + Convection + Evaporative Cooling.

In general, virtually any adaptation for maintaining body temperature (morphological, physiological or behavioral) can be related to one or more of the components in this equation. In reality, the complexity of heat budget equations is far greater when various factors are included, e.g., insulation, wind speed, humidity, etc. This section of Chapter 6 concludes with some examples of heat budgets, which include some of these factors.

The second section details the effect of temperature on metabolism as well as the difference between acclimation and acclimatization. The primary focus in this section is to describe the van't Hoff or Q_{10} rule (a doubling of metabolic rate for each 10°C rise in temperature). This section provides basic information for understanding contrasts drawn between poikilotherms and homeotherms in later sections of Chapter 6.

Students will find the next two sections regarding plant and animal responses to temperature intriguing. To the extent possible, these responses should be related to the organism's manipulation of specific elements of the heat budget. Plants experience a highly variable thermal environment because of their immobility; consequently, they respond differently to temperature than animals. Plants balance heat inputs by radiation, convection and evapotranspiration. In general, many aspects of plant shape can be related to the need for temperature regulation. Similarly, dormancy in plants is often a response to prolonged cold. Some plants also exhibit thermogenesis, usually as a way of attracting pollinators. In general, however, plants are more limited in their responses to temperature than animals; thus, they possess various mechanisms for dealing with heat stress and cold stress.

The section regarding animals begins with an explanation of the three general classifications of temperature regulation among animals: poikilotherms, homeotherms and heterotherms. Students should understand that homeotherms maintain body temperature by means of endothermy and that poikilotherms maintain body temperature through ectothermy. This does not necessarily mean that the terms homeotherm and endotherm, nor poikilotherm and ectotherm, are synonymous. The advantages/disadvantages of both endothermy and ectothermy should be emphasized.

For poikilotherms, ectothermy has the advantage of limiting metabolic costs associated with maintaining body temperature through oxidative metabolism; hence, less food is required and more energy can be allocated to biomass production. Additionally, these organisms are not limited to a minimum size. On the other hand, limitation of activity to those times in which temperatures are sufficiently warm is a significant disadvantage for poikilotherms. The differences between small versus large poikilotherms, aquatic versus terrestrial poikilotherms and in acclimation versus acclimatization are presented in this section. Additionally, various mechanisms for survival and/or temperature regulation are presented and should be considered in the classroom, e.g., heliothermism, proportional control, burrowing, supercooling, and diapause.

By producing heat through metabolism, homeotherms are less constrained by thermal environments. Nonetheless, the main disadvantage of homeothermy is a higher food requirement to maintain metabolism. For homeotherms, body size is also an important consideration, since metabolic rate is proportional to the 0.75 power of body mass (i.e., metabolic rate varies inversely with body weight). Consequently, below a minimum size, metabolic rate reaches a maximum beyond which the energy requirements cannot be met. Homeotherms use a variety of physiological and morphological means to regulate temperature, e.g., counter-current exchange, retes, thermal windows, gular fluttering, insulation (fat, fur, or feathers), shivering, thermogenesis, sweating, panting, etc.

The organisms classified as heterotherms exhibit characteristics of both endothermy and ectothermy. A variety of heterotherms are described in the text with emphasis on important considerations about thermal balance in these organisms. For example, flying insects are essentially ectothermic when at rest and are endothermic while in flight. Similarly, true hibernators and endotherms that enter daily torpor can be considered as heterotherms because of the body temperature decrease that occurs during these quiescent periods. In some cases, ectotherms use localized retes to warm muscles or brain tissues, e.g., tuna and swordfish.

The concluding section of Chapter 6 summarizes the salient information about the influence of temperature on distributions of plants and animals. The central idea is that species reside within upper and lower limits of temperature tolerance. Many species are generally restricted by the lowest critical temperature of their life cycle. Numerous examples are provided demonstrating the interaction of temperature and organismal distributional patterns. For plants, a practical suggestion would be to include some of the information from Chapters 4 (Climate) and 5 (Moisture) for an integrated consideration of the factors controlling plant distributions.

TOPICS FOR DISCUSSION

1. Would you classify most plants, bacteria and fungi as being poikilotherms? Why?

2. What mechanisms may have led to the evolution of endothermy? How did thermogenesis evolve in some plants?

3. Given that a bear exhibits only minimal temperature depression during dormancy, does it really hibernate?

4. Relate the idea of migration to the thermal balance (heat budget) equation. Using this equation as a basis, explain why some organisms must migrate seasonally.

5. Identify which element(s) of the heat budget equation is/are affected by the following adaptations and behaviors:

gular fluttering	sweating
burrowing	heliothermism
insulation	endothermy
counter-current exchange	thermal window

6. Some leaves of plants are too small in size to minimize overheating. What other environmental factors affect leaf size? Present an overall discussion of the factors that interact to determine leaf shape and size.

7. Given that C_4 plants are more productive and have higher water use efficiencies than C_3 plants, why haven't C_4 plants outcompeted and replaced the evolutionarily older C_3 plants?

8. Since ectotherms are free from the constraints of small body size, what are the advantages of being small?

9. On the basis of what you know about temperature regulation, what arguments can you formulate both for and against the idea that dinosaurs were endothermic?

10. Describe the various types of dormancy for both plants and animals and how these relate to thermal balance for these species.

11. Use the following information about an organism and its heat budget to answer the questions below:

A. total radiant energy input into an organisms = 1.4 kcal/cm² of body surface/min;
B. convection = 0.3 kcal/cm²/min;
C. evaporative cooling = 0.4 kcal/cm²/min;
D. conduction = 0.2 kcal/cm²/min.

How much radiation is this organism losing? If the optimal output for this organism is 0.8 kcal/cm²/min, what would be some appropriate behavioral or physiological responses by this organism?

LECTURE OUTLINE

I. Introduction - Thermal Constraints on Organisms

II. Heat Budgets - Thermal Balance

    A. Inputs
        1. Insolation
        2. Diffuse Radiation
        3. Reflection from Environmental Objects
        4. Metabolism
    B. Outputs
        1. Radiation
        2. Convection
        3. Conduction
        4. Evaporative Cooling
    C. Other Models of Thermal Balance
        1. Animals - Insulation
        2. Plants - Wind Speed & Boundary Layer

LECTURE OUTLINE CONTINUED:

III. Responses to Temperature

    A. Influence of Temperature on Metabolism
    B. Plants
        1. Coping with Heat Stress
        2. Coping with Cold Stress
        3. Thermogenesis
    C. Animals
        1. Poikilotherms
            a. Advantages - Disadvantages
            b. Active Temperature Range
            c. Influence of Body Size
            d. Aquatic Poikilotherms
        2. Homeotherms
            a. Advantages - Disadvantages
            b. Influence of Body Size
        3. Heterotherms
            a. Flying Poikilotherms
            b. Dormancy in Homeotherms
            c. Localized Heat Production
        4. Examples of Temperature Responses in Animals
            a. Behavioral Responses
                i. heliothermism
                ii. proportional control
                iii. burrowing
                iv. migration
            b. Anatomical Structures
                i. thermal windows
                ii. retes
                iii. insulation
            c. Physiological Responses
                i. thermogenesis
                ii. hyperthermia
                iii. acclimation and acclimatization
                iv. dormancy
                v. shivering
                vi. supercooling
                vii. sweating and panting

IV. Influence of Temperature on Species' Distributions

    A. Limits of Tolerance
        1. Variation with Life Stage
        2. Minimum Critical Temperature
    B. Example Distributions

RESOURCE MATERIALS

Some of the references presented for earlier chapters are also relevant here because they include information, if not entire chapters, about temperature and/or heat budgets, e.g., Larcher (1980), Barbour et al. (1987), Rosenberg (1983), Schmidt-Nielsen (1990). In addition to these, many works exist regarding thermal regulation; for example, the *Journal of Thermal Biology* is dedicated to presenting papers describing the mechanisms by which temperature affects living organisms.

For insects, the two articles in *Advances in Insect Physiology* are very useful (Wilmer 1982; Casey, 1985). Casey considers both endothermy and ectothermy in insects and the consequences of flight and various avenues of heat exchange. Wilmer provides case histories of both thermal and water balance for various insect groups (water balance information is appropriate to Chapter 5).

Barbour, M., J. H. Burk, and W.D. Pitts. 1987. *Terrestrial Plant Ecology* (2nd Edition). The Benjamin/Cummings Publishing Company, Inc., Menlo Park, CA.

Casey, T.M. 1985. "Thermoregulation and heat exchange." *Advances in Insect Physiology* 20:119-146.

French, A.R. 1988. "The pattern of mammalian hibernation." *American Scientist* 76:568-577.

Heinrich, B. (editor). 1981. *Insect Thermoregulation*. John Wiley and Sons, Inc., New York.

Hill, R.W. and G.A. Wyse. 1989. *Animal Physiology* (2nd Edition). Harper and Row, New York.

Larcher, W. 1980. *Physiological Plant Ecology*. Springer-Verlag, New York.

Rosenberg, N.J. 1983. *Microclimate: The Biological Environment* (2nd Edition). Wiley and Sons, Inc., New York.

Schmidt-Nielsen, K. 1990. *Animal Physiology: Adaptation and Environment* (4th Edition). Cambridge University Press, New York.

Willmer, P.G. 1982. "Microclimate and environmental physiology of insects." *Advances in Insect Physiology* 16:1-57.

CHAPTER 7

LIGHT AND BIOLOGICAL CYCLES

CHAPTER SUMMARY AND ORGANIZATION

Chapter 7 describes the many ways that light influences the activity and distribution of plants and animals. The chapter demonstrates that any consideration of the ecological role of light involves more than just a consideration of photosynthesis. Many facets of Chapter 7, however, might be combined in the classroom setting with consideration of other topics, e.g., heat budgets/temperature, stratification in lakes (Chapter 15), photosynthesis (Chapter 10), succession (Chapter 30), and others. On the other hand, the chapter provides a useful backdrop for chapters and topics to come.

The chapter begins with an overview of the nature of light and the degree to which light is reflected, absorbed or transmitted in terrestrial and aquatic habitats. Some of this information would undoubtedly be relevant to discussions of photosynthesis (Chapter 10), e.g., the topics of photosynthetically active radiation, leaf area index (LAI) and compensation intensity.

The discussion in the second section of Chapter 7, regarding plant adaptations to light, is primarily in the context of shade tolerant versus shade intolerant plant species. Various adaptations and responses of shade tolerant species are described. This theme predominates in a discussion of both terrestrial habitats and submersed aquatic habitats, the latter of which constitute dim habitats beneath the water's surface. This section concludes with a brief discussion of the effects of ultraviolet radiation on plants.

The following section about photoperiodism forms a substantial portion of Chapter 7. The average student has probably never given much consideration to the ideas of circadian rhythms and biological clocks, but they should because humans, and presumably all living creatures, are very much affected by biological clocks, which are generally synchronized with Earth's photoperiod. Students should understand that the circadian rhythm is actually an endogenous cycle of about 24 hours that is under genetic control (see Aronson et al., 1994, in the **Resource Materials**). This internal cycle (or clock) is set by external environmental cues. The most reliable "timesetter" for species, on both a daily and seasonal basis, is the length of day and night. To one degree or another, light influences many biological events, such as daily and seasonal activity (nocturnal versus diurnal or dormant versus active), migration, food gathering, mating, diapause, initiation of flowering, flower opening and closing, and the list continues. The effects of light are considered from two points of view: First, the time-keeping mechanisms that relate to internal physiological clocks are considered. Information about free-running cycles, entrainment, phase shift, and models of biological clocks are presented. Second, the role of critical daylength and many examples of its effects are described. This section concludes with a brief summary about the possible existence of circannual clocks.

In some habitats, activities are not synchronized by just light-dark cycles. The next component of Chapter 7 illustrates that tidal and lunar rhythms may dictate the activities of marine species, especially intertidal organisms. Similarly, seasonal cycles in the tropics are not necessarily a function of light-dark cycles; rather, rainfall events are the important environmental cues.

The final section on seasonality builds upon the information presented about biological clocks. The primary idea is that seasonal biological activities are often influenced by the interaction of a variety of factors, primarily the interaction of light, temperature, and moisture. Depending on latitude, the importance of each of these factors varies, e.g., seasonal events in the tropics are more or less synchronized with wet and dry periods. Floristic seasonal phenomena are detailed with discussions of flowering, vegetative growth, and allocation of biomass and energy. Similarly, seasonal activities of animals are presented, including the subjects of diapause, migration, reproduction, molting, nesting, and decomposition.

TOPICS FOR DISCUSSION

1. Describe the aspects of human behavior that are under the control of an endogenous biological clock. Explain the relationship between phase shift and jet lag in humans.

2. In a forest setting, try to identify, observe and characterize some shade tolerant plants and shade intolerant plants. Are there any common characteristics of shade tolerant plants? How would tolerance of shading affect succession of plant species?

3. Describe the ecological significance of circadian rhythms. How and why have these rhythms evolved?

4. Why aren't most seasonal cycles controlled by a single environmental cue, such as light?

5. Cave animals either live in complete darkness or they only leave caves at night (i.e., some bats). Speculate on the operation of biological clocks in these organisms.

LECTURE OUTLINE

I. Nature of Light

    A. Photons
        1. Wave and Particle Characteristics
        2. Wavelength
        3. Electromagnetic Radiation
    B. Reflection of Light
        1. Terrestrial Habitats
            a. Albedo
            b. Leaf Surfaces
        2. Aquatic Habitats
    C. Absorption of Light
        1. Atmospheric Absorption
        2. Absorption by Leaves
        3. Absorption in Water
            a. Influence of Turbidity
            b. Extinction Coefficient
    D. Transmission of Light

II. Responses to Light

    A. Shade Tolerant and Intolerant Plant Species
        1. Morphology
        2. Photosynthetic and Respiratory Characteristics
        3. Light Compensation and Light Saturation

LECTURE OUTLINE CONTINUED:

    B. Periodicity: Environmental Cues and Endogenous Cycles
        1. Circadian Rhythms
            a. Significance and Examples
            b. Environmental Cues and Endogenous Cycles
                i. free-running cycle
                ii. entrainment
                iii. phase shifts
        2. Biological Clocks
            a. Two-oscillator Model
            b. Hourglass Model
        3. Daylength

III. Seasonality

    A. Interaction of Light, Temperature and Precipitation
    B. Examples of Seasonal Plant Phenomena
        1. Seed Germination
        2. Flowering and Fruiting
        3. Biomass Allocation
        4. Deciduous Plants
    C. Examples of Seasonal Animal Phenomena
        1. Dormancy
        2. Migration
        3. Reproduction

IV. Other Periodic Cycles

    A. Circannual Clocks
    B. Tidal and Lunar Cycles
    C. Wet-Dry Periods in the Tropics

RESOURCE MATERIALS

A vast literature exists regarding the influences of light on biological processes. In addition to the references provided in the text of Chapter 7, the following citations can provide useful background information for preparation of lectures and provision of additional readings for students. Bainbridge and Evans (1966) and Evans et al. (1975) give excellent overviews of the significance of light as an ecological factor and perhaps provide the best context for presenting information about light in an introductory ecology class. Riklis (1988) is an extensive book with detailed papers on the effects of UV light and circadian rhythms (papers are included on circadian rhythms in mole rats, pigeons, green finches and siskins). For more information describing circadian rhythms and biological clocks, Daan and Gwinner (1989) and Winfree (1987) can be quite useful. Similarly, Attridge (1990) considers circadian rhythms in plants. There is also a journal, *The Journal of Biological Rhythms*, dedicated to the study of biological clocks and various types of rhythms, including circadian rhythms. For information about plants, Attridge (1990), Whatley and Whatley (1980) and Evans et al. (1988) are recommended. The emphasis of these books is primarily physiological (but with ecological implications); however, Whatley and Whatley (1980) do include a chapter regarding light as an ecosystem factor. Attridge (1990) also presents an introductory chapter regarding the nature of light. Finally, Herring et al. (1990) and Kirk (1983) are good sources for consideration of light in aquatic environments. Each book includes information about the physics of light in aquatic media, as well as information about ecological phenomena. Kirk (1983), in particular, provides a chapter regarding the ecological strategies of aquatic plants in relation to variation in light intensity and spectral quality. Herring et al. (1990)

include sections on animal behavior and bioluminescence. Recent articles regarding the genetic aspects of circadian rhythms have been written by Aronson et al. (1994) and Page (1994).

Aronson, B.D., K.A. Johnson, J.J. Loros and J.C. Dunlap. 1994. "Negative feedback defining a circadian clock: autoregulation of the clock gene frequency." *Science* 263:1578-1584.

Attridge, T.H. 1990. *Light and Plant Responses: A Study of Plant Photophysiology and the Natural Environment*. Edward Arnold, London.

Bainbridge, R. and G.C. Evans. 1966. "Light as an Ecological Factor." *Symposium of the British Ecological Society, Volume 6*. Blackwell Scientific Publishing, Oxford.

Barbour, M., J. H. Burke, and W.D. Pitts. 1987. *Terrestrial Plant Ecology* (2nd Edition). The Benjamin/Cummings Publishing Company, Inc., Menlo Park, CA.

Daan, S. and E. Gwinner (editors). 1989. *Biological Clocks and Environmental Time*. The Guilford Press, London.

Evans, G.C., R. Bainbridge, and O. Rackham. 1975. "Light as an Ecological Factor II." *Symposium of the British Ecological Society, Volume 16*. Blackwell Scientific Publishing, Oxford.

Evans, J.R., S. von Caemmerer, and W.W. Adams III (editors). 1988. "Ecology of Photosynthesis in Sun and Shade." Reprints from *Australian Journal of Plant Physiology*, Volume 15, Numbers 1 and 2, CSIRO, Australia.

Herring, P.U., A.K. Campbell, M. Whitfield, and L. Maddock (editors). 1990. *Light and Life in the Sea*. Cambridge University Press, NY.

Larcher, W. 1980. *Physiological Plant Ecology*. Springer-Verlag, New York.

Kirk, J.T.O. 1983. *Light and Photosynthesis in Aquatic Ecosystems*. Cambridge University Press, Cambridge.

Page, T.L. 1994. "Time is the essence: Molecular analysis of the biological clock." *Science* 263:1570-1572.

Riklis, E. 1988. "Photobiology: The Science and its Applications." *Proceedings of the Tenth International Congress on Photobiology*. Plenum Press, New York.

Whatley, J.M. and F.R. Whatley. 1980. "Light and Plant Life". *The Institute of Biology's Studies in Biology, Number 124*. The Camelot Press, Ltd., Southampton.

Winfree, A.T. 1987. "The Timing of Biological Clocks." *Scientific American Library Series, Number 19*. W.H. Freeman, New York.

Wolken, J.J. 1986. *Light and Life Processes*. Van Nostrand Reinhold Co., New York.

CHAPTER 8

NUTRIENTS

CHAPTER SUMMARY AND ORGANIZATION

This chapter outlines many ecological aspects of nutrition in plants and animals. While the chapter does emphasize the nutrition of individuals and the various localized effects and cycles of nutrients, it also provides a sound basis for understanding global biogeochemical cycles of nutrients with significant gaseous forms (Chapter 12).

The chapter begins by providing a brief introduction to 20 essential macronutrients and micronutrients. Table 8.1 describes the biological role of each these elements. Whether or not the specific details and symptoms of deficiency should be stressed for each nutrient in the classroom is a decision that should be left to the instructor, but the elements which are of particular biological importance and/or cycle globally certainly should be considered. Students should understand that not only are nutrient deficiencies a problem for organisms, but excessive concentrations of essential elements can have adverse consequence for life (particularly the micronutrients). This idea can be related back to Shelford's Law of Tolerance, in which organisms live within a zone determined by minimum and maximum amounts of a substance needed for existence. Students will certainly have a hard time remembering the essential elements, but an old mnemonic device used to help with remembering 16 of the elements required by plants is as follows:

C HOPKNS CaFe Mg / B Mn CuZn MoCl.

These letters represent the abbreviations for 16 elements and the expression can be read as "C. Hopkins Cafe - management by my cousin Moclue." Of course, this is a little silly and its doubtful anyone is named "Moclue," but students will remember this phrase; thus, they will remember 16 of the elements required by all living organisms. The slash mark in the expression represents the division between macro- and micronutrients. Also, iodine is not included in the phrase (i.e., HOPKNS) because it isn't required by plants. Hydrogen is not described in Chapter 8.

After the initial introduction to essential elements, the second section of the chapter describes some important components of local nutrient cycles. Most of the examples given in this section are for temperate forests, but the information can be related to other terrestrial systems. Most facets of local cycles can be presented effectively in the context of nutrient budgets. The primary sources of nutrient input into ecosystems are decomposition, mineral weathering, and atmospheric deposition. While soil is the main direct source of nutrients for most plants, atmospheric deposition (both wet and dry) can be important in nutritionally deficient soils. Regardless of the nutrient source, plants seem well-adapted for keeping nutrients within the system. Decomposition, stemflow, throughfall, and leaching all act to concentrate elements around plants (particularly trees). Within a system, not all of the molecules of an element cycle at the same rate; some are stored for various lengths of time (usually as organic detritus), while others cycle quickly (short-term cycle). The mix of plant life history strategies in a forest seems to ensure the existence of both short-term and long-term nutrient pools. In summary, this section should impress upon students that in any system nutrients must be available, yet conserved, for the stability of the system.

Chapter 8 ends with specific examples of how plants and animals, respectively, are affected by nutrient regimes. The effects of nutrient availability on plant density, distribution, competition, and many other facets of plant existence are discussed. In particular, the differences between calcifuges, calcicoles, and neutrophilus plants are detailed in the context of pH effects, calcium availability and metal toxicity. In addition, the adaptations of plants to extreme conditions of excessive element concentrations are described, e.g., serpentine soils, metal contaminated soils, saline soils (inhabited by halophytes), and atmospheric pollutants.

For consumers, the ultimate source of nutrition is plant material. The most observable effect of nutrients on consumers usually relates to nutrient deficiencies. In general, sodium, calcium and magnesium are known to affect the distribution, behavior, fitness and, possibly, the cyclic population patterns of some animals. The text presents a variety of examples illustrating these phenomena. In general, demonstrable effects of nutrition are more obvious for herbivores than carnivores because by eating plant material, herbivores are consuming a food that may be very different in terms of the proportions and quantities of elements needed, i.e., plants assimilate and store elements in proportions and masses different from animals. Carnivores, in contrast, consume a diet containing nutrients more typical of animal needs. Consequently, these generalizations, coupled with the usual high availability of plant materials, make herbivores more dependent on food quality, whereas carnivores are more dependent on food quantity.

TOPICS FOR DISCUSSION

1. Some plants are carnivorous, others plants form unique mutualistic relationships with microbes, e.g., relationships with *Rhizobium* (a nitrogen-fixing bacteria) or with various fungi that form mycorrhizae protruding from the roots of the plant. How do these unique characteristics relate to the need for nutrients in these plants?

2. Contrast the nutrient storage in a tropical forest with a temperate forest. Why are some tropical soils deficient in nutrients despite supporting luxuriant growth of plants? Why are decomposition rates often very fast in tropical systems, and how does this relate to the nutrient requirements of the plants?

3. How does Shelford's Law of Tolerance relate to nutrient requirements in organisms?

4. Arrange the 10 most abundant nutrients in plants and/or animals in order of decreasing concentration. How does this order relate to the abundance of the same nutrients in various kinds of ecosystems?

5. How is nutrient transfer along a food chain coupled to the need for energy? On the other hand, how is the utilization or capture of energy related to adequate nutrition in plants or animals?

6. Describe the nutrient quality of plants at various times during a growing season. Why do many herbivores prefer young plant material? How would the production of digestion-reducing substances by plants interfere with the nutrition of various herbivores?

7. The eventual release of nutrients from plant litter depends in part upon degradation by microbes, which also must have their nutritional needs met. If plant detritus cannot supply the nutritional needs of heterotrophic microbes, then what is the source of nutrition for these microbes? Why does the detrital complex often show a short-term increase in concentrations of nitrogen and phosphorus?

LECTURE OUTLINE

The flexibility regarding lecture organization for information about nutrients is considerable. The outline below presents one organization of the primary material in Chapter 8. Another approach, however, would be to organize lectures around a theme of nutrient cycling/biogeochemical cycling, in which a presentation about nutrients and their ecological significance is followed by a consideration of local nutrient cycles and global cycles (Chapter 12). This organization also could be structured to include information regarding soil chemistry and storage, as presented in Chapter 9 (Soils), and processes unique to specific nutrients, e.g., nitrogen and sulfur.

I. Overview of Important Nutrients

    A. Macronutrients
    B. Micronutrients

II. Local Nutrient Budgets and Cycles

    A. Nutrient Inputs
        1. Precipitation - Wetfall and Dryfall
            a. Interception
            b. Throughfall
            c. Stemflow
        2. Weathering
    B. Nutrient Outputs
        1. Leaching
        2. Erosion
    C. Simple Input - Output Model

III. Specific Effects of Nutrients on Plant and Animal Species

    A. Plant Species
        1. Calcicoles, Calcifuges and Neutrophilus plants
        2. Serpentine Soils
        3. Salinity
        4. Environmental Pollutants
    B. Animal Species
        1. General Considerations
            a. Dependency on Plants
            b. Herbivores versus Carnivores
        2. Sodium Deficiency

RESOURCE MATERIALS

References are provided below from *Scientific American*, *The Annual Review of Ecology and Systematics* and *Advances in Ecological Research*. Each of these citations is a review article appropriate for an introductory ecology course. While some of these papers are older, they are still relevant for considerations of nutrient cycling in ecosystems and provide a wealth of examples and considerations. In some instances, these papers also will be relevant to the following chapter regarding soils (Chapter 9), e.g., Coleman et al. (1983) and Marschner (1988). Some of the papers consider specific components of nutrient cycles or elemental effects on plants. For example, Parker (1983) provides a detailed analysis of factors affecting stemflow and throughfall with the inclusion of a discussion of elements leached by and dissolved in throughfall. Similarly, Antonovics et al. (1972) detail the ecology and evolution of heavy metal tolerance in plants, and Procter (1975) thoroughly describes the ecological significance of serpentine soils. Coleman et al. (1983) detail many of the biological aspects of the C, N, P and S cycles in soils. Binkley and Richter (1987) provide a forested ecosystem context for evaluating H^+ sources (including acid precipitation), H^+ pools and H^+ sinks, including effects of C, N, P, S, and Ca cycles on H^+. For an overview of mineral cycling in plants, Marschner (1988) describes functions of nutrients, symptoms of deficiencies, adaptations of plants to adverse soil conditions, and more. Newman (1988) provides a fascinating examination of potential links between plants via mycorrhizae, in which the functioning of these links in the transport of substances between plants is considered.

Antonovics, J., A.D. Bradshaw, and R.G. Turner. 1971. "Heavy metal tolerance in plants" *Advances in Ecological Research* 7:2-85.

Bormann, F.H. and G.E. Likens. 1970. "The nutrient cycles of an ecosystem." *Scientific American* 223:92-101.

Coleman, D.C., C.P.P. Reid, and C.V. Cole. 1983. "Biological strategies of nutrient cycling in soil systems." *Advances in Ecological Research* 13:1-55.

Deevey, E.S. 1970. " Mineral cycles." *Scientific American* 223:148-158.

Jordan, C. F. and J.R. Kline. 1972. "Mineral cycling: Some basic concepts and their application in a tropical rain forest." *Annual Review of Ecology and Systematics* 3:33-50.

Jordan, C.F. 1985. *Nutrient Cycling in Tropical Forest Ecosystems: Principles and Their Application in Management and Conservation.* John Wiley and Sons, Inc., New York.

Marschner, Horst. 1988. Mineral Nutrition of Higher Plants. Academic Press, New York.

Parker, G.G. 1983. "Throughfall and stemflow in the forest nutrient cycle." *Advances in Ecological Research* 13:57-120.

Pomeroy, L.R. 1970. "The strategy of mineral cycling." *Annual Review of Ecology and Systematics* 1:171-190.

Proctor, J. and S.R.J. Woodell. 1975. "The ecology of serpentine soils." *Advances in Ecological Research* 9:255-366.

CHAPTER 9

SOILS

CHAPTER SUMMARY AND ORGANIZATION

The average biology student is probably unaware of the importance of soil as an essential resource. Directly or indirectly, all living organisms depend on soil. An ecology course is probably one of the few places in which a student of biology will be exposed to soil as an integral component of ecosystems; thus, information in Chapter 9 should be considered essential to most courses in general ecology. As a system consisting of mineral, gaseous, and aqueous components, the soil system provides numerous functions. It serves as a site for nutrient storage and delivery, nutrient transformation, water storage, waste deposition and processing, root anchorage for plants, burrowing and nesting, escape, etc. In addition to these roles, soils and soil formation processes can be related to variation in topography, climate and vegetation.

The first two sections of Chapter 9 describe the problem of precisely trying to define soil and the organization of a soil profile. At a minimum, soil should be emphasized as being composed of both inorganic constituents and organic by-products. Typically, the soil components are organized into layers (horizons), which are formed by differential processes of deposition, leaching and weathering. Of the five layers that can be identified (O, A, B, E and C), the O and A horizons support the greatest biological activity. Later in Chapter 9, the ecological significance of the O horizon and the role of organisms in this layer during soil formation are discussed.

In the next four sections of Chapter 9, soil properties, soil biota and soil genesis/development are considered. Because soils of widely varying characteristics generally can be found in close proximity to one another in any region, this information can be reinforced by having students in the lab characterize soil samples from various field sites. Better yet, introduce students to soil and soil properties in the field, where any topographic or biotic effect on soil type will be readily apparent. The important soil properties are color, texture, moisture, and chemistry, each of which is discussed in the second and third sections of Chapter 9. These properties should be considered thoroughly in the classroom or field because they infer much about (1) the environmental conditions under which the soil is developing (e.g., anerobiosis); (2) water availability to plants; (3) nutrient availability to plants; (4) soil quality as habitat for plants and animals. It will be virtually impossible to discuss each of these independently from one another in the classroom. For example, texture affects the permeability of the soil and the water storage capacity of the soil, both of which play a role in determining oxygen status, which, in turn, affects soil color. Similarly, the clay texture class is particularly important in the formation of clay-humus micelles, which are the key to nutrient retention and availability. This latter consideration should include the relationship between soil pH and cation exchange capacity. This discussion also could be augmented with relevant information regarding relationships between acid rain, soil buffering capacity and cation exchange capacity in soils.

Generally, soil scientists consider soil genesis separately from soil formation, but the distinction is sometimes vague. Soil genesis, the breakdown of parent rock material into smaller particles, includes the physical, chemical and biological degradation of rock material, whereas soil formation deals more with the development of soil with distinctive characteristics. The appearance of a soil at any point in time is dependent on many important interactions between a variety of variables, such as climate, biological activity, parent rock material and elapsed time, all of which are discussed in detail. Of these factors, it seems only natural in an ecology class to emphasize the pervasive biotic influence during soil development. The text stresses the deposition and decomposition of organic matter during soil formation, particularly the O horizon, in its discussion of biotic influences during soil genesis. Furthermore, the section entitled *The Living Soil* provides a detailed summary of the diversity and function of organisms residing within soil. Students should realize that a reciprocal relationship exists between the soil flora and fauna and organic debris (litter) on and in the soil. Specifically, the text details the importance of soil organisms to the development of specific O horizon types called mull, moder and mor.

In more of a regional context, students also should understand that the interaction of vegetation and climate over broad regions results in the identification of five major soil development processes: podzolization, laterization, calcification, salinization and gleyization. Each of these processes is described and placed into the appropriate regional and climatic perspective. The key point here is that great spatial and temporal variation exist among soils. On a local scale, spatial variation may relate to topography, hence the term, toposequence. On the other hand, chronosequences represent soils of different ages that originated from the same parent material.

Chapter 9 concludes with a brief presentation of mismanaged soils. To this point, the information about soil profiles, genesis and specific developmental processes relate to undisturbed soils. Many soils, however, have been greatly disturbed by human activities. In addition to direct disturbance resulting from construction and agricultural activities, soil erosion can occur when the vegetative component of an ecosystem is disturbed or removed. The text defines sheet erosion, rill erosion and gully erosion.

In summary, Chapter 9 should not be overlooked in an introductory ecology course. As already emphasized, soil plays a fundamental role in all ecosystems. For some animals, their entire life is spent within the confines of soil. From a pedagogical viewpoint, chapter 9 provides an absolutely essential base for understanding the soil compartment of biogeochemical cycles (Chapter 12). Furthermore, a basic understanding of soil characteristics is needed to facilitate comparisons among ecosystems, particularly terrestrial systems.

TOPICS FOR DISCUSSION

1. How does the cultivation of land affect the organic content of soil, the soil profile, the bulk density of soil, the organisms that might reside in the soil, and soil erosion?

2. The transition from open water in a lake to an upland terrestrial system is very gradual. Describe the changes you might see in soil properties if you were to begin to sample at the water's edge of a lake and continue sampling at increasing distances away from this lake. To what would you attribute these changes?

3. Much emphasis is placed on the effect of acid precipitation on aquatic systems, e.g., the Adirondack Lakes. How does soil chemistry alter the impact of acid deposition in aquatic systems? What are some of the effects of acid deposition on soils?

4. Why do heavy, clay soils generally have fewer invertebrate inhabitants in relation to more loosely textured soils?

LECTURE OUTLINE

There are so many ways to organize information in ecology that the options can almost be overwhelming. You may find it useful to present some of the information in Chapter 9 in different contexts. Within the chapter itself, one might find it more convenient to begin with a consideration of how a soil comes to exist (i.e., soil formation), and to use this as a base for understanding specific soil characteristics, such as the soil profile and other soil properties. In a broader context, some of the information on soil organisms could be included in a discussion of detrital energetic pathways or the role of decomposition in nutrient cycling. Similarly, one might find it convenient to incorporate the specific processes of podzolization, laterization, calcification, and gleyization into lectures on the comparative ecology of various ecoregions, such as temperate forests, tropical forests, prairies, tundras, etc. Obviously then, the following outline is only one possible organizational scheme for a lecture regarding soils and their ecological significance.

I. Soils: What are they?

    A. Essential Components
       1. Mineral Material
       2. Organic Material
    B. Phases
       1. Solid Phase
       2. Liquid Phase
       3. Gaseous Phase
    C. Ecological Roles
       1. Nutrient Storage and Delivery
       2. Water Storage and Delivery
       3. Biological Significance
              a. Media for Plant Anchorage
              b. Site for Microbial Processing
                 i.   waste breakdown
                 ii. elemental transformation
              c. Habitat for Soil-Dwelling Animals

II. Soil Genesis and Formation

    A. Weathering Processes
       1. Physical Processes
       2. Chemical Processes
    B. Important Influences Affecting Soil Genesis
       1. Parent Rock Material
       2. Climate
       3. Biota
       4. Topography
       5. Time

    C. Important Soil Development Processes
       1. Podzolization
       2. Laterization
       3. Calcification
       4. Salinization
       5. Gleyization

III. Disturbed Soils

    A. Physical Disruption of Soil Horizons
    B. Soil Compaction
    C. Soil Erosion

RESOURCE MATERIALS

Donahue et al. (1983), Ross (1989) and Steila and Pond (1989) all present many facets of soil science. These books do differ in the approach used by the authors. Donahue et al. (1983) more or less compartmentalize many important aspects of soil, e.g., sections exist on soil water, soil organic matter, soil biology, soil fertility, soil development, physical properties of soils and chemical properties of soils, among others. On the other hand, in a very informative book, Steila and Pond (1989) consider the origins and classification of soils with emphasis on temporal and spatial variation. Individual chapters are presented on each of the soil orders according to the U.S. Comprehensive Soil Classification System. Ross (1989), in contrast, considers the various processes affecting or occurring in soils. For example, she describes the processes characteristic of saturated soils, weathering processes, degradation of organic matter, solute transfer, plant rooting zones, etc. Ross also includes an environmental perspective in her book by including chapters on the effects of cultivation on soils, namely ploughing, fertilization and pesticide application. Jenny's work (1980) is an older, but excellent resource. In addition to these references, Coleman (1983) is relevant and considers the physics, chemistry and biology of soil-related processes in an ecosystem context.

Barbour, M., J. H. Burk, and W.D. Pitts. 1987. *Terrestrial Plant Ecology* (2nd Edition). The Benjamin/Cummings Publishing Company, Inc., Menlo Park, CA.

Coleman, D.C., C.P.P. Reid, and C.V. Cole. 1983. "Biological strategies of nutrient cycling in soil systems." *Advances in Ecological Research* 13:1-55.

Donahue, R.L., R.W. Miller, and J.C. Shickluna. 1983. *An Introduction to Soils and Plant Growth* (5th Edition). Prentice-Hall, Inc., NJ.

Jenny, H. 1980. "The Soil Resource: Origin and Behavior." *Ecological Studies, Volume 37*. Springer-Verlag, New York.

Ross, S. 1989. *Soil Processes - A Systematic Approach*. Routledge, Chapman and Hall, Inc., New York.

Steila, D. and T.E. Pond. 1989. *The Geography of Soils: Formation, Distribution, and Management* (2nd Edition). Rowman and Littlefield Publishers, Inc., MD.

CHAPTER 10

THE ECOSYSTEM

CHAPTER SUMMARY AND ORGANIZATION

Chapter 10 contains information absolutely essential to all introductory ecology courses, since ecosystem theory represents a cornerstone of ecology. This chapter is also important because of the information it contains relevant to energy flow and nutrient processing, which are discussed in later chapters. One organizational approach that may be taken to introduce ecosystems in the classroom would be to examine the fundamental components of all systems. Namely, systems have (1) various components, (2) inputs and outputs, and (3) a specific function, usually; ecosystems are no exception. The organization of Chapter 10 lends itself well to this approach. At the simplest level, the components and inputs/outputs can be either biotic (living) or abiotic (non-living). In other words, the major inputs are energy (various forms), elements/molecules (various forms) and organisms/propagules; similarly, the outputs are energy, elements and organisms/propagules, but not usually identical to the inputs. The components can be organized in various ways, but include functional groups of organisms: autotrophs, heterotrophs and detritivores. The question of what is (are) the function(s) of ecosystems is a complex one. Usually, however, ecosystems function to process the energy and nutrients through which life is supported. Thus, the importance of understanding energy flow and nutrient cycling becomes evident.

After an initial overview of the lithosphere, hydrosphere and biosphere, Chapter 10 begins with a synopsis of the evolution of the ecosystem concept. While certainly important, the extent to which one may want to emphasize the historical overview in classroom lectures is a matter of personal choice. As an alternative organizational scheme, some of this information could be combined with the information in Chapter 1 to provide an integrated discussion/lecture of the emergence of ecology as a discipline. One might consider this approach for the first one or two lectures of an introductory ecological course.

The critical information to stress in the classroom relates to the similarities between ecosystems and other kinds of systems. In fact, for beginning students, it is often instructive to compare an ecosystem to a system with which students might be familiar, e.g., a microcomputer system. The microcomputer has inputs (energy and information/data), outputs (altered energy and processed information/data), components (CRT, keyboard, CPU, printer, etc.) and has the function of processing information/data. With this system one can draw analogies to ecosystems and their characteristics. Of course, extreme care must be taken to emphasize that ecosystems do not have finite boundaries, which many other systems possess.

The second section of Chapter 10 is subdivided into sequential discussions of photosynthesis and decomposition. This material fits well into an introductory lecture about ecosystems. However, the information may fit equally well into lectures regarding energy flow and/or nutrient cycling. For example, photosynthesis represents the initial energetic transformation as energy is captured by plants and made available to consumers. By comparison, decomposition is the terminal stage of energy flow, as well as a key process in all nutrient cycles. Despite the organizational scheme to be utilized in lectures, both photosynthesis and decomposition need to be considered in detail, because they represent two essential processes common to nearly all ecosystems. One exception occurs at great depths near volcanic vents in oceans. In these unique ecosystems, photosynthesis is not important; rather, chemosynthetic autotrophs use hydrogen sulfide (H_2S) released from the Earth's crust via the volcanic vent.

The section on photosynthesis provides an excellent consideration of C_3, C_4 and CAM photosynthetic pathways, topics often overlooked in other ecology textbooks. These pathways represent important physiological adaptations to moisture and temperature stress and should be discussed thoroughly in the classroom. If information about moisture and temperature have already been discussed in previous lectures, this is a good time to revisit some of that information to emphasize how the physiology of plants has responded to these critical abiotic parameters. Furthermore, one theme for any ecological class should be the nature of interrelationships between all components of ecosystems, a point which can be made effectively by using photosynthetic mechanisms.

The concluding section on decomposition often is overlooked in the classroom by instructors of ecology. This is unfortunate because decay of organic matter is as essential as the production of organic matter. From an energy processing standpoint, more energy fixed by photosynthesis is processed by detrital food chains than grazing food chains. Yet, most texts consider grazing food chains in great detail while providing only minimal coverage of decomposition and detrital processing. Further, many aquatic systems (e.g., shaded, heterotrophic streams) are driven by the decay of organic matter entering the system rather than by photosynthetic processes. Chapter 10 provides basic information regarding decomposition, all of which should be covered in a lecture format. Instructors are encouraged to expand on this information by supplementing lectures from other sources and providing additional reading for students (see **Resource Materials**).

In the broader context of an ecology course, Chapter 10 provides a firm basis for structuring various laboratory activities. For example, the discussion of photosynthesis provides descriptions of photosynthetic efficiencies and leaf area index. Laboratories easily can be designed to assess the conversion of solar energy into plant biomass. Solar energy can be estimated by the use of energy recording devices (radiometers or pyranometers) or may be simply obtained from published tables of solar radiation at various latitudes. Biomass can be harvested and converted to energy units via bomb calorimetry or the use of published values. Similarly, decomposition and the succession of detritivores can be studied by using the simple litterbag technique (see **Resource Materials** for information regarding litterbags). This technique involves the enclosure of detritus in mesh bags, which allow the entry and exit of detritivores, as well as the loss of organic matter over time.

TOPICS FOR DISCUSSION

1. Elevated levels of CO_2 often are considered from the standpoint of potential changes in photosynthetic rates and how C_3 and C_4 plants will be differentially affected. In a slightly different context, how will elevated CO_2 levels perhaps affect the nutrient content and nutritional quality of plant material for herbivores? (See Grodzinski, 1992 in **Resource Materials**.)

2. If global warming does occur, to what extent will the decay of organic matter be affected?

3. Individual organisms represent very special biological systems. What are the differences and similarities between ecosystems and organisms? In Chapter 1 of the textbook, the work of F.E. Clements is described, in which communities/ecosystems are considered as super-organisms. How does this analogy hold-up or break-down in the context of system theory?

4. How does a change in leaf area index relate to light availability and limitation for plants?

5. Why is energy flow and usage so critical to our understanding of the operation of natural systems? How do human consumption and utilization of energy relate to energy capture and utilization by ecosystems (the context can be either historical or contemporary)?

LECTURE OUTLINE

The outline below closely follows the textbook--with two major exceptions. The outline recommends coverage of limitations on photosynthesis and factors affecting rates of organic matter decay. These topics are not covered in Chapter 10, but fit well into a unified, classroom consideration of photosynthesis and decomposition. Of course, these two processes could be combined with information in Chapter 11 (Ecological Energetics), since critical components of energy flow in natural systems. Plan on spending at least one full lecture each on photosynthesis and decomposition. Thus, the following outline probably will take about three one-hour lectures, particularly if historical aspects are provided about the origin of ecosystem theory.

I. Ecosystem Concept: Origin and Definition

II. Essential Components of Systems/Ecosystems

    A. Inputs/Outputs
          1. Abiotic
          2. Biotic
    B. Biotic Components
          1. Autotrophs
          2. Heterotrophs
          3. Decomposers
    C. Two Essential Processes
          1. Photosynthesis
                a. Dark Reactions
                      i. C_3
                      ii. C_4
                      iii. CAM
                b. Photosynthetic Efficiencies
                c. Limitations on Photosynthesis
                      i.   light
                      ii.   temperature
                      iii. water
                      iv.   nutrients
          2. Decomposition
                a. Stages of Decay
                b. Important Organisms
                      i.   biotic succession
                      ii.   functional groups
                c. Determinants of Decay
                      i.   nutrient regimes (internal and external)
                      ii.   temperature
                      iii. pH
                      iv.   oxygen
                      v.   nature of organic matter
                      vi.   biotic influences

RESOURCE MATERIALS

The citations in the textbook regarding the historical and introductory aspects of ecosystems are more than adequate reading for additional information. Thus, the following works relate primarily to photosynthesis and decomposition. Further information regarding photosynthesis can be found in virtually any introductory botany textbook (e.g., Northington and Goodin, 1984, or Rost et al., 1984,) or plant physiology textbook (e.g., Salisbury and Ross, 1978). Both Osmund (1989) and Mooney and Field (1989) help place the process of photosynthesis into a proper global perspective. In addition to these references, intense interest is now focused on photosynthetic responses to elevated CO_2 levels in Earth's atmosphere, as evidenced by a collection of papers in *Science, Bioscience, Scientific American* and the *Annual Review* of *Ecology and Systematics* (see below).

In contrast to literature regarding photosynthesis, information about decomposition is generally fragmented among myriad research and review papers. Some of the more salient review papers are provided below, especially for aquatic ecosystems (Polunin, 1984; Webster and Benfield, 1986). Weider and Lang (1982) provide an excellent overview, particularly to the use of litterbags, which can be used in designing a laboratory to examine decay processes.

Bazzaz, F.A. 1990. "The response of natural ecosystems to the rising global CO_2 levels." *Annual Review of Ecology and Systematics* 21:167-196.

Chazdon, R.L., R.W. Pearcy. 1991. "The importance of sunflecks for forest understory plants." *Bioscience* 41:760-766.

Culotta, E. 1995. "Will plants profit from high CO_2?" *Science* 268:654-656.

Grodzinski, B. 1992. "Plant nutrition and growth regulation by CO_2 enrichment." *Bioscience* 42:517-525.

Lamarche, V.C., D.A. Graybill, H.C. Fritts, and M.R. Rose. 1984. "Increasing atmospheric carbon dioxide: Tree ring evidence for growth enhancement in natural vegetation." *Science* 225:1019-1021.

Mooney, H.A. and C.B. Field. 1989. "Photosynthesis and plant productivity - scaling to the biosphere." In: *Photosynthesis* (W.R. Boiggs, editor) - Proceedings from the C.S. French Symposium on photosynthesis held in Stanford, California, July 17-23, 1988. (pp. 19-44.) Alan R. Liss, Inc., New York.

Mooney, H.A., B.G. Drake, R.J. Luxmoore, W.C. Oechel, and L.F. Pitelka. 1991. "Predicting ecosystem responses to elevated CO_2 concentrations." *Bioscience* 41:96-104.

Moore, P.D. 1994. "High hopes for C_4 plants." *Science* 367:322-323.

Northington, D.K. and J.R. Goodin. 1984. *The Botanical World.* Times Mirror/Mosby College Publishing, St. Louis, MO.

Olson, J. 1963. "Energy storage and the balance of producers and decomposers in ecological systems." *Ecology* 44:322-331.

Osmond, C.B. 1989. "Photosynthesis from the molecule of the biosphere: a challenge for integration." In: *Photosynthesis* (W.R. Boiggs, editor) - Proceedings from the C.S. French Symposium on photosynthesis held in Stanford, California, July 17-23, 1988. (pp. 5-17.) Alan R. Liss, Inc., New York.

Polunin, N.V.C. 1984. "The decomposition of emergent macrophytes in freshwater." *Advances in Ecological Research* 14:115-116.

Rost, T.L., M.G. Barbour, R. M. Thornton, T.E. Weier, and C. R. Stocking. 1984. *Botany - A Brief Introduction to Plant Biology.* John Wiley and Sons, Inc., New York.

Salisbury, F.B. and C.W. Ross. 1978. *Plant Physiology.* Wadsworth Publishing Company, Belmont, California.

Webster, J.R. and E.F. Benfield. 1986. "Vascular plant breakdown in freshwater ecosystems." *Annual Review of Ecology and Systematics* 17:567-594.

Weider, R.K. and G.E. Lang. 1982. "A critique of the analytical methods used in examining decomposition data obtained from litterbags." *Ecology* 63:1636-1642.

White, R.M. 1990. "The great climate debate." *Scientific American* 263:36-43.

CHAPTER 11

ECOSYSTEM ENERGETICS

CHAPTER SUMMARY AND ORGANIZATION

One of the primary functional facets of ecosystems is to process energy; hence, Chapter 11 details energy flow and usage in ecosystems. Many aspects of Chapters 7 (light) and 10 (sections on photosynthesis and decomposition) are relevant here and should be reviewed or re-emphasized at appropriate points during lectures.

Before ecological energetics can be discussed, students must have a basic understanding of energy, its various forms and the constraints on energy usage. Thus, Chapter 11 begins with an introductory overview of relevant definitions and principles regarding the nature of energy. Some instructors may want to review or integrate information from Chapter 7 regarding the nature of light into this lecture. At the very least, students should thoroughly understand the first and second law of thermodynamics. In this context, emphasize in the classroom that ecological systems are open, steady-state systems in which entropy is offset by the continual input of free energy. With the exception of the hydrothermal vents previously mentioned in Chapter 10, ecosystems are driven, directly or indirectly, by energy from the sun after capture by photosynthetic organisms (e.g., vascular and non-vascular plants, algae and some bacteria).

As an extension of the process of energy capture by plants (see Chapter 10 regarding photosynthesis), the second section of Chapter 11 details storage and utilization of energy by plants. Three aspects of this section need to be stressed in lectures. First, the definitions and concepts related to energy accumulation by plants should be discussed. Students should thoroughly understand the ideas of biomass, standing crop, net primary production and gross primary production.

Second, lectures should emphasize that plants exhibit many strategies for energy allocation and utilization, all of which maximize survival and reproductive success. Allocation of energy (and other resources) to roots and shoots is a fundamental consideration; root:shoot ratios can be related to availability of water (Chapter 5) and nutrients (Chapter 8); further, the plant communities of various ecosystems often have characteristic root:shoot ratios. Other important points to be made in lecture relate to energetic allocations to supporting tissues (e.g., stems and trunks), reproductive structures, and seasonal allocation patterns. Various organizational approaches for understanding life history strategies will be presented later in Chapter 20.

Third, students should understand that levels of primary production vary immensely among ecosystems, between ecosystems of the same type, and within the same ecosystem from year to year. Comparisons should be drawn between ecosystems of high and low productivity. In general, the productivity of terrestrial ecosystems is most influenced by temperature and precipitation patterns. On a smaller scale, temporal and spatial variation in productivity can be related to nutrient availability, grazing pressure, outbreaks of plant disease or pest infestation, fire and growing season length.

The discussion of primary productivity concludes with a description of the methodology used to estimate production rates in aquatic and terrestrial ecosystems. Close examination of these methods by students should help with their understanding of the difference between net and gross primary production. The information in this section may be integrated effectively into a laboratory unit in which field production measurements are conducted.

After the presentation of primary production, the latter half of Chapter 11 is divided into four major sections, all of which deal with the utilization of energy by heterotrophs and the transfer of energy through the ecosystem. These sections begin with a consideration of secondary production, which is a measure of production by heterotrophs. The approach in this section is to utilize an organismal energy budget to define secondary production and demonstrate the possible fates of ingested energy. Additionally, the energy budget is used to present an initial consideration of some efficiencies (ratios) of energy utilization by individual organisms. Certainly, an appropriate lecture scheme would be to integrate a consideration of efficiencies of energy transfer at the organismal level and ecosystem level. Efficiencies of energy flow through an ecosystem are obviously a manifestation of the efficiencies of energy consumption and usage by single organisms (see textbook section regarding "models of energy flow").

After the discussion of secondary production, food chains are presented. Food chain components (e.g., herbivores, carnivores, omnivores and decomposers) and types of food are separately considered. These sections differ from most textbooks in that the food chain components are not presented in the context of trophic levels. Trophic levels are considered later in the chapter in the discussion of ecological pyramids. This separation allows one thoroughly to consider the implications of different diet types in the function and evolution of consumers; thus, the components of food chains are discussed from the point of view of dietary differences and requirements. For herbivores, food is plentiful, but the diet is constrained by low protein levels and the relative indigestibility of cellulose in plant materials. In contrast, carnivores are usually not constrained by diet quality; rather, their major constraint is related to obtaining sufficient amounts of food through the capture of elusive prey. Hence, we see adaptations in herbivores for increasing digestion and assimilation of plant materials. Conversely, adaptations in carnivores generally are related to increasing the success of prey capture.

The major food chains are separated into detrital and grazing chains (with a brief consideration of supplementary food chains). Several key points need to be made with regard to the separation of detrital and grazing food chains. First, separation of the two energetic pathways occurs primarily at the first linkage between trophic levels, in which primary production is either consumed by herbivores or deposited as dead organic matter (a smaller proportion of organic matter deposition obviously results from consumers). Second, only a very small proportion of primary production enters the grazing food chain via herbivores (usually less than 10%); thus, most of the energy fixed in photosynthesis enters the detrital pathway. Third, the two food chains are interconnected; a significant proportion of the biomass of consumers in the detrital food chain is utilized by consumers in the grazing food chain.

From a functional point of view, each step of a food chain represents a trophic level (feeding level). When one considers individual species, however, this hierarchy seems somewhat artificial given that many organisms feed at more than one trophic level. Furthermore, detrital processors are not encompassed by the trophic concept. Nonetheless, the trophic level concept is a useful approach to understanding energetic transfers in ecosystems, and trophic levels are used in a number of organizational schemes, e.g., pyramids of numbers, biomass and energy are organized on the basis of trophic levels representing specific layers in each pyramid. Of these three pyramid types, the most instructive in an energetic context is the pyramid of energy, which is based on energy capture, storage and transfer. Energy pyramids clearly demonstrate the flow and diminishment of energy along a grazing food chain.

Despite the amount of lecture time that will be dedicated to the food chains, trophic levels and energy pyramids, students must recognize at some point the degree of artificiality that surrounds the terminology and organizational schemes to depict energy passage in an ecosystem. As emphasized in Chapter 11's discussion of food webs, trophic relationships are virtually never linear and involve the meshing of numerous food chains and organisms feeding at various levels and in various food chains, i.e., complex food webs dominate in nature. In this context, application of trophic relationships to food webs becomes very difficult. One attempt to overcome such difficulties relates to the idea of trophospecies, an organizational scheme in which species are grouped together on the basis of their similarity as prey or predators. Lectures regarding food webs also should emphasize that web complexity is often related to the productivity of an ecosystem. Insure that students understand that complexity does not arise from increased food chain length. The complexity arises from the support of more species and therefore an increase in the number of linkages in the food web.

The final paragraph in the textbook's presentation of food webs describes cascading food chains, whereby removal of a top predator creates a cascade of effects down through the food chain. This is a recent, but very controversial, idea in aquatic ecology. If this idea is to be discussed in lecture, instructors should read some of the papers presented in the **Resource Materials** for further background (e.g., Demelo et al., 1992).

Instructors probably will want to discuss ecological efficiencies (ratios of energy inputs and outputs within or between trophic levels). These efficiencies can be used to describe important aspects about the function of different organisms. For example, relevant comparisons of some ecological efficiencies can be drawn between poikilotherms and homeotherms or between herbivores and carnivores. Table 11.2 summarizes various ecological efficiencies. No matter how the material is organized for lectures, students should understand the significance of the exploitation efficiency, the assimilation efficiency, and the production efficiency.

Chapter 11 contains a wealth of information essential to any course in introductory ecology. Many of the concepts presented are long-standing ecological cornerstones. To depart somewhat from the organization of the textbook, I recommend that the salient features of ecological energetics are best explained in the classroom by taking an energy budget approach from the outset with the grazing food chain used as the primary model. First, detail energy passage along the grazing food chain and emphasize that unused energy from each trophic level enters the detrital food chain, which will be considered subsequently. The autotrophs should be included as a key component of any food chain. Initially, describe the differences between primary and secondary production, as well as the differences between net and gross production. After students have a general understanding of energetic linkages along the food chain, the energy budget approach can be used to consider appropriate ecological efficiencies and ecological pyramids. The efficiencies can be used for incorporation of important information about specific characteristics of organisms within each trophic level, e.g., adaptations for energy capture can be related to the efficiency of solar energy utilization in plants, energy allocation in plants can be linked to their net production efficiency of plants, adaptations to a plant diet by herbivores can be related to the assimilation efficiency, etc. After presentation of the grazing food chain, important comparisons can be made to the detrital food chain and both of these food chains can be used to emphasize food web theory. Please refer to the recommended lecture outline for additional details.

TOPICS FOR DISCUSSION

1. Ecological efficiencies often tell an ecologist much about the biology of an organism, i.e., these efficiencies are not just numbers, they are biologically meaningful. For example, what would a low net production efficiency for a plant suggest? Would you expect trees and grass to have similar net production efficiencies? How would you expect the production efficiency to differ between endotherms and ectotherms, between active and inactive organisms, between small and large organisms? What does the assimilation efficiency suggest about the quality of an organism's diet?

2. The biomass of phytoplankton in aquatic systems is sometimes less than the biomass of zooplankton preying on the phytoplankton. How can this be?

3. In which ecosystems would you expect food web complexity to be highest? Given what you know about factors governing primary productivity in ecosystems, would you expect positive correlations between food web complexity and any prevailing climatic factors? If so, which factors?

4. Given human nature, high efficiencies often are equated with being "better" than low efficiencies. Why doesn't this apply to organismal biology? For example, endotherms are quite successful, yet their production efficiency is low compared to ectotherms. Similarly, the net production efficiency of a tree is low compared to that of an aquatic plant, but both plants do very well in their respective habitats. Why?

5. It has been asserted that one function of an ecosystem is to process energy. Justify that viewpoint on the basis of the information presented in Chapter 11.

LECTURE OUTLINE

The following outline incorporates the material from Chapter 11; however, material from Chapter 10 with regard to photosynthesis and decomposition is also included. Instructors will need to make their own decision about whether to include this information here or elsewhere. See the lecture outline for Chapter 10 as an alternative. The following outline probably will require at least three one-hour lectures.

I. The Nature of Energy

    A. Definitions and Units of Measurement
    B. States of Energy - Kinetic and Potential
    C. Forms of Energy
        1. Chemical
        2. Atomic
        3. Electrical
        4. Mechanical
        5. Electromagnetic
    D. Laws of Thermodynamics

II. The Grazing Food Chain

    A. Components
        1. Autotrophs
            a. Energy capture (see Chapter 10, photosynthesis)
            b. Gross and net primary production
        2. Heterotrophs
            a. Types
                i.   herbivores
                ii. carnivores
                ii. omnivores
                iv. detritivores
            b. Important Steps in Energy Transfer
                i.   consumption
                ii. assimilation
    B. Important Ecological Efficiencies
        1. Within Trophic Levels
            a. Autotrophs
                i.   Lindeman efficiency (see Chapter 10)
                ii. net production efficiency (NPP/GPP)

LECTURE OUTLINE CONTINUED:

           b. Heterotrophs
              i.   exploitation efficiency
              ii.   assimilation efficiency
              iii. production efficiency
      2. Between Trophic Levels
              i.   production efficiency
              ii.   assimilation efficiency
  C. Ecological Pyramids
      1. Pyramids of Numbers
      2. Pyramids of Biomass
      3. Pyramids of Energy

III. The Detrital Food Chain
  A. Decay of Organic Material (See Chapter 10)
  B. Component Organisms
  C. Significance

IV. Food Webs
  A. Interaction of Detrital and Grazing Food Chains
  B. Food Web Complexity

RESOURCE MATERIALS

Given that the flow of energy through ecosystems is one of the unifying principles in ecology, any other introductory textbook to ecology will provide additional information to complement the material presented in Chapter 11. Ricklefs (1990) provides a detailed, up-to-date treatment with an overall organization similar to Smith's *Ecology and Field Biology*. See Colinvaux (1993) for a slightly different point of view than that given here. The idea of cascading food chains is briefly presented in the latter sections of Chapter 11. Some researchers vehemently defend this idea while other scientists are a bit more skeptical. For further information regarding this division, refer to the evaluation of application of top-down trophic control in aquatic systems by Demelo et al. (1992) and to Carpenter and Kitchell's response to that paper (Carpenter and Kitchell, 1992) . More general reading is also provided in Spencer et al. (1991).

Carpenter, S.R. and J.F. Kitchell. 1992. "Trophic cascade and biomanipulation: Interface of research and management - A reply to the comment by DeMelo et al". *Limnology and Oceanography* 37:208-213.

Colinvaux, P. 1993. *Ecology 2*. John Wiley and Sons. New York.

Demelo, R., R. France and D. J. McQueen. 1992. "Biomanipulation: Hit or myth?" *Limnology and Oceanography* 37:192-207.

Ricklefs, R. 1992. *Ecology* (3rd Edition). W.H. Freeman and Company, New York.

Spencer, C.N., B. R. McClelland, and J.A. Stanford. 1991. "Shrimp stocking, salmon collapse, and eagle displacement - Cascading interactions in the food web of a large aquatic system." *Bioscience* 41:14-21.

CHAPTER 12

BIOGEOCHEMICAL CYCLES

CHAPTER SUMMARY AND ORGANIZATION

This chapter provides a basic overview to elemental cycles. The chapter is organized on the basis of the two cycle types: gaseous and sedimentary. Oxygen, carbon and nitrogen cycles are presented as examples of gaseous cycles. Sedimentary cycles are represented by sulfur, phosphorus and lead. Special attention will need to be given to the sulfur cycle, since it exhibits characteristics of both gaseous cycles and sedimentary cycles. For example, gaseous sulfur-containing molecules are common (SO_2 and H_2S), and sulfur cycles globally because of these gases. Throughout Chapter 12, attention is given to environmental consequences associated with human interference in biogeochemical cycles. For each elemental cycle, the extent and significance of human impacts on these elemental cycles are discussed. Additionally, Chapter 12 concludes with three sections regarding contemporary environmental problems related to the introduction of serious, human-generated contaminants into ecosystems and food chains; sequentially, the consequences of acid deposition, chlorinated hydrocarbons and radionuclides are presented in these sections.

The introduction to Chapter 12 emphasizes that energy flow and elemental cycling are closely coupled, with each limiting the other. In addition, biogeochemical cycles, gaseous cycles and sedimentary cycles are defined. Some instructors may want to augment this introduction with a presentation/review of the hydrologic cycle (Chapter 5) and basic information regarding essential elements (Chapter 8). In particular, the hydrologic cycle is fundamental to understanding most elemental cycles.

The first two gaseous cycles discussed are the oxygen and carbon cycles. Diagrams of the cycles are presented in Figures 12.2 and 12.5, respectively. In each case, the important pools and transformations are presented. The focus of the oxygen presentation relates to the production and consumption of ozone, with emphasis on human-induced problems, e.g., introduction of gaseous oxides and chlorofluorocarbons. The section regarding carbon cycling is more extensive describing both local and global circulation patterns. Instructors are encouraged to couple the oxygen and carbon cycles with the flow of energy in their lectures. Not only are several carbon compounds important pools of oxygen (CO_2, carbohydrates, etc.), the capture and utilization of energy involve a series of oxidation-reduction reactions. Photosynthesis is the reduction of oxidized carbon (CO_2) with the release of molecular oxygen. Conversely, aerobic, heterotrophic activity involves the oxidation of reduced carbon, the consumption of oxygen and the release of CO_2. The flow of energy through food chains is essentially the movement of reduced, organic carbon compounds along the food chain. The dynamic state of atmospheric CO_2 in relation to photosynthetic - respiratory processes is illustrated on both a daily and seasonal basis (see Figures 12.6 and 12.8). The discussion of the carbon cycle concludes with a detailed consideration of the greenhouse effect and human impacts on global carbon cycling. In particular, a number of consequences of elevated CO_2 and potential atmospheric warming are presented, which students probably have not considered, e.g., ecosystem shifts, impacts on ecosystem productivity, change in growing seasons, and many others.

In comparison to the oxygen and carbon cycles, students will find the nitrogen cycle more complex; consequently, this cycle is presented in detail. First, the important processes of nitrogen transformation are presented. Nitrogen fixation, denitrification, ammonification and nitrification are thoroughly discussed. Students should understand these processes prior to presentation of the nitrogen cycle. Figure 12.16 places each transformation in its proper context in the overall global nitrogen cycle. Given the complexity of nitrogen transformations and cycling, instructors also may find it beneficial to present an overview of the various forms and oxidation states of nitrogen. Nitrogen can occur in seven valence states (see Söderlund and Rosswall, 1982 - **Resource Materials**), and a consideration of the oxidation of reduced compounds is biologically and ecologically relevant. Denitrification involves the reduction of nitrites and nitrates, whereby the molecules function as electron acceptors to release energy from reduced carbon compounds. In contrast, nitrification involves the release of energy by the oxidation of reduced (relatively speaking) nitrogen compounds. Given that the biological basis of ecological energetics is oxidation and reduction, the nitrogen and carbon cycles are perfect examples to demonstrate the coupling of energy processing and elemental processing.

The effect of human activity on nitrogen availability and cycling concludes Chapter 12's presentation of nitrogen. A portion of this section can be related back to ozone production and degradation. Other important aspects of this discussion relate to the implications of nitrogen enrichment in ecosystems, particularly forested systems.

The section regarding sedimentary cycles begins with the sulfur cycle. Figure 12.18 provides a detailed diagram of sulfur circulation. Three points should be emphasized. First, note the sedimentary and gaseous phases of sulfur. Truly sedimentary cycles involve two phases for elements, salt solution and rock; clearly, sulfur is an exception. Second, emphasize that various human activities, particularly those related to energy consumption, result in major inputs of sulfur to the atmosphere. The human influence on atmospheric sulfur is relevant to the presentation of acid deposition. Finally, discuss the significance of oxidation and reduction processes in sulfur cycling. Again, it may be useful to describe the important forms of sulfur and their valence states. In this context, the reasons for sulfur oxidation and reduction by microbes are more obvious to students. For example, sulfur reduction occurs in anaerobic habitats and is essentially analogous to denitrification, with oxidized sulfur molecules acting as electron receptors during carbon oxidation.

The second element to be presented in the context of sedimentary cycles is phosphorus. Relative to the other cycles presented in Chapter 12, phosphorus cycling is rather straightforward. Phosphorus cycling is simplified by the absence of significant gaseous compounds. Students should understand that, although phosphorus is absolutely essential to all living organisms, it is a comparatively rare element and especially limiting in aquatic systems. In aquatic systems, phosphorus occurs in four fractions: particulate, inorganic, and low and high molecular weight organic compounds. Phosphorus enrichment in aquatic ecosystems has led to spectacular increases in algal productivity and the phenomenon of cultural eutrophication. For a detailed presentation of phosphorus in aquatic systems, refer to Wetzel (1983 - see **Resource Materials**).

The final discussion of sedimentary cycles involves a consideration of lead. Lead is an excellent example of a heavy metal. In minute quantities, heavy metals have always cycled in ecosystems. Because of human activity, however, these metals now occur in high concentrations in many localities and in the tissues of many organisms, including humans. If instructors so desire, a wealth of information regarding many heavy metals can be found in Moore and Ramamoorthy (1984 - see **Resource Materials**).

The discussion of lead cycling introduces an environmental theme that carries throughout the rest of Chapter 12. In the final three sections, the ecological impacts of three human-induced environmental problems are described. Initially, acid deposition is discussed. In poorly-buffered aquatic systems, acidity can be related to aluminum toxicity, increased mortality for various aquatic species and declines in reproduction. In terrestrial systems, acidity is believed to be related to declines in cation exchange capacity, inhibition of fungi and bacteria, decreases in elemental transformations, defoliation of conifers and vegetational shifts in forests. Next, environmental contamination by chlorinated hydrocarbons (primarily DDT and PCBs) is considered. These compounds have become distributed globally, even in areas never receiving direct inputs, and concentrations often magnify in animal tissues. These compounds can accumulate to toxic levels, but they also cause numerous metabolic problems even at sublethal levels. In addition, magnification of DDT in fish tissues has rendered the fish of some localities unfit for human consumption. The final section of Chapter 12 describes the release and significance of radioactive contaminants in Earth's environment. These compounds can exist for thousands of years, during which time they enter food chains with debilitating effects, particularly to genetic systems. The textbook provides examples of radioactive contamination from both terrestrial and aquatic environments, with emphasis on uptake by plant and animal species.

In conclusion, instructors should keep in mind that two possible themes can be developed from this chapter. The first theme relates to ecosystem function which, as mentioned in Chapter 10 of this manual, includes the processing/cycling of nutrients. Of course, nutrient cycling and energy utilization are closely coupled and are two facets of life support. The second theme that can be developed relates to humankind's blatant, and sometimes naive, impact on this planet's chemistry. Most students will be aware of these environmental concerns only through their reading of popular articles, and this probably will be their first opportunity for a serious consideration of environmental contaminants.

TOPICS FOR DISCUSSION

1. Instruct each student to consider his or her place in some of the nutrient cycles. Ask them to think about the food they consume, the products they utilize, the atmosphere they breathe. What specific physiological and structural adaptations do humans have for extracting nutrients from their food?

2. How does the harvesting of plants and animals from specific ecosystems affect nutrient cycles? Consider not only loss of nutrients, but the effects that are manifested by replenishing nutrients in the form of fertilizers. Ask students to think about the consequences of nutrient redistribution. Where do nutrients harvested in the form of biomass ultimately get deposited?

3. How are nutrient cycles interwoven? For example, does carbon processing affect the oxygen cycle, and vice versa? In aquatic environments, how do sulfur transformations affect phosphorus processing?

4. Will acid precipitation affect the cycling of nutrients in ecosystems? When and how? Under what circumstances may impacts be maximized?

5. What are some possible effects of atmospheric CO_2 enrichment on plant nutrient content, plant decay and nutrient processing? (See White, 1990; Mooney et al., 1991; Bazzaz and Fajer, 1992 - **Resource Materials.**)

LECTURE OUTLINE

Students with poor backgrounds in chemistry may find information about biogeochemical cycles difficult; consequently, an approach in which the important forms, pools and transformation are sequentially considered before presentation of an overall cycle may be desirable. Consideration of the environmental consequences associated with humankind's impact on the various cycles can be discussed sequentially, as presented in the textbook, or may be considered collectively after all relevant biogeochemical cycles have been discussed.

I. Introduction

    A. Definitions and Types of Cycles
    B. Significance of Hydrologic Cycle (see Chapter 5)

II. Gaseous Cycles

    A. Oxygen Cycle
        1. Important Transformations
            a. Photosynthesis - Respiration
            b. Photodissociation of Water
        2. Global Pools of Oxygen
        3. Relation to Carbon Cycle
        4. Human-related Intrusions - Ozone Depletion
    B. Carbon Cycle
        1. Relevant Forms
            a. Natural Buffer System - H_2CO_3 - HCO_3 - CO_3
            b. Atmospheric CO_2
            c. Pools of Organic Carbon
        2. Important Transformations
        3. Regional and Global Fluxes and the Carbon Cycle
        4. Human-related Intrusions
    C. Nitrogen Cycle
        1. Relevant Forms and Oxidation States
        2. Important Transformations
            a. Nitrogen Fixation
            b. Denitrification
            c. Ammonification
            d. Nitrification
        3. Global Nitrogen Cycle
        4. Human-related Intrusions

III. Sedimentary Cycles

    A. Sulfur Cycle
        1. Relevant Forms and Oxidation States
        2. Important Transformations
            a. Sulfur Oxidation
                i.   photosynthetic bacteria
                ii.   chemosynthetic bacteria
            b. Sulfur Reduction
        3. Global Sulfur Cycle
        4. Human-related Intrusions
    B. Phosphorus Cycle
        1. Relevant Forms
        2. Aquatic Phosphorus Cycles
        3. Human-related Intrusions
    C. Consideration of Heavy Metals - Lead

IV. Cycling and Consequences of Some Human-produced Products

    A. Acid Deposition
    B. Chlorinated Hydrocarbons
    C. Radionuclides

RESOURCE MATERIALS

Because of the environmental implications associated with various biogeochemical cycles, a wealth of information is available as additional resources. For example, *The Handbook of Environmental Chemistry* series encompasses many volumes containing chapters dedicated to specific cycles (e.g., see Söderland and Rosswall, 1982, for the nitrogen cycle or Zehnder, 1982, for the carbon cycle). Information regarding phosphorus cycling, particularly in aquatic systems, is extensive; for a general overview consult Stumm (1973). Detailed reviews of phosphorus cycling also may be found in Pierrou (1976), Richey (1983) Stumm and Morgan (1981) and Wetzel (1983). Coleman et al. (1983), Deevey (1970) and Jordan and Kline (1972) are three older references that provide overviews to mineral cycles suitable for an introductory ecology course. For general information regarding significant environmental problems, Moore and Ramamoorthy provide separate chapters for each heavy metal and the problems associated with their release and cycling. Laws (1993) provides an introductory text regarding pollutants in aquatic systems with chapters on eutrophication, pesticides, acid deposition, and radioactivity, as well as many others. In addition, three papers provide excellent reviews of the predicted effects of atmospheric CO_2 enrichment: White (1990); Mooney et al. (1991); Bazzaz and Fajer (1992).

Bazzaz, F.A. and E.D. Fajer. 1992. "Plant life in a CO_2-rich world." *Scientific American* 266:68-74.

Coleman, D.C., C.P.P. Reid, and C.V. Cole. 1983. " Biological strategies of nutrient cycling in soil systems." *Advances in Ecological Research* 13:1-55.

Deevey, E.S. 1970. "Mineral Cycles." *Scientific American* 223:148-158.

Fukai, R. and Y. Yokoyama 1982. "Natural Radionuclides in the Environment." In: *The Handbook of Environmental Chemistry*, Volume 1, part b: *The Natural Environment and the Biogeochemical Cycles*. O. Hutzinger (editor). Springer-Verlag, New York.

Jordan, C.F. and J. R. Kline. 1972. "Mineral Cycling: Some basic concepts and their application in a tropical rain forest." *Annual Review of Ecology and Systematics* 3:33-50.

Laws, E. 1993. *Aquatic Pollution - An Introductory Text* (2nd edition). John Wiley and Sons, Inc., New York.

Mooney, H.A., B.G. Drake, R.J. Luxmoore, W.C. Oechel, and L.F. Pitelka. 1991. "Predicting ecosystem responses to elevated CO_2 concentrations." *Bioscience* 41:96-104.

Moore, J.W. and S. Ramamoorthy. 1984. *Heavy Metals in Natural Waters*. Springer-Verlag, New York.

Pierrous, U. 1976. "The global phosphorus cycle." In: *Nitrogen, Phosphorus and Sulfur - Global Cycles*. B.H. Svensson and R. Söderlund (editors). Scientific Committee on Problems of the Environment (SCOPE) Report 7. *Ecological Bulletin* (Stockholm) 22:75-88.

Richey, J.E. 1983. "The phosphorus cycle." In: *The Major Biochemical Cycles and Their Interactions*. B. Bolin and R.B. Cook (editors). Scientific Committee on Problems of the Environment. (SCOPE) Report 21. John Wiley and Sons, Inc., New York.

Stumm, W. 1973. "The acceleration of the hydrogeochemical cycling of phosphorus." *Water Research* 7:131-144.

Stumm, W. and J.J. Morgan. 1981. *Aquatic Chemistry - An Introduction Emphasizing Chemical Equilibria in Natural Waters*. John Wiley and Sons, Inc., New York.

Söderland, R. and T. Rosswall. 1982. "The Nitrogen Cycles." In: *The Handbook of Environmental Chemistry*, Volume 1, part b: *The Natural Environment and the Biogeochemical Cycles*. O. Hutzinger (editor). Springer-Verlag, New York.

Wetzel, R.G. 1983. *Limnology*. Saunder's College Publishing. New York.

White, R.M. 1990. "The great climate debate." *Scientific American* 263:36-43.

Zehnder, A.J.B. 1982. "The Carbon Cycle." In: *The Handbook of Environmental Chemistry*, Volume 1, part b: *The Natural Environment and the Biogeochemical Cycles*. O. Hutzinger (editor). Springer-Verlag, New York.

CHAPTER 13

GRASSLAND TO TUNDRA

CHAPTER SUMMARY AND ORGANIZATION

Chapters 13 and 14 provide detailed descriptions about specific terrestrial ecosystems. These chapters build on the generalized aspects of ecosystem processes introduced in Chapters 9 - 12, but with more specific examples and application of the ideas already introduced. Description of ecosystems should not be overlooked in an introductory class. From a number of different points of view, this information is important. For example, one could argue that the science of ecology evolved from plant geography, in which the distribution of plant life-forms were eventually mapped and related to patterns of temperature and moisture, i.e., what we now consider ecosystem types. Similarly, most ecologists eventually specialize, to one degree or another, in the study of organisms or processes in specific ecosystem types. Furthermore, many of today's environmental concerns relate to the destruction and over-exploitation of ecosystems. In the United States, examples include concern about excessive logging in old-growth forests, over-grazing of natural grasslands, loss of wetlands, contamination of aquatic ecosystems, the effects of acid deposition on forests, and the list continues. Thus, students must come away from an introductory ecology class with a rudimentary understanding of the organization and function of the major ecosystems. Toward this end, Chapter 13 introduces and describes five specific terrestrial ecosystems, some of which are subdivided into smaller systems. The organization of each ecosystem description is generally based on three categories: (1) an overview of ecosystem characteristics, (2) a description of the physical structure intrinsic to the ecosystem, and (3) an explanation of ecosystem function, which focuses on the processing of energy and nutrients.

The systems presented are grasslands, savannas, shrublands, deserts and tundra, respectively. The common link among these systems is a generally low-moisture regime for at least part of the year. Despite a wide range of average annual temperatures across these five ecosystems, precipitation is typically much lower than in forested systems. Low moisture restricts the vertical structure, biomass and productivity regimes of these plant communities; thus, the physiognomy of these systems tends to be simple in comparison to forests.

The first three systems presented, grasslands, savannas and shrublands, are the hardest to characterize. Many varieties of each of the ecosystems occur; for example, eight grasslands are introduced in the text encompassing cultivated systems, systems dominated by perennial grasses and systems dominated by annual grasses. Some of the systems may even be considered desert grasslands. In the case of savannas, which possess both grass and woody components, definitions are not precise and these systems range from being almost exclusively grassland to almost shrubby. Four different shrublands are characterized, including shrublands that are transitory successional stages preceding forests.

Despite the immense variety within each of these ecosystem types, one can envision a gradient encompassing grasslands, savannas (with grass and woody components), and shrublands, the latter of which are dominated by woody species (grasses also can be common). Commonalities among these ecosystems can be emphasized in lectures. These systems are somewhat xeric, have low productivity, are adapted to fire and typically have high root:shoot ratios. Similarly, animals are superbly adapted for life in these systems, and adaptations for grazing, speed, camouflage, keen vision and burrowing are readily evident.

With regard to deserts, low rainfall and high evaporative potential are universal. The text describes the geographic and topographic features that contribute to arid climatic conditions and characterizes rainfall ranges for semideserts, true deserts and extreme deserts. The influence of low and infrequent water availability should be the focal point of classroom lectures/discussion. Aridity influences most aspects of desert life, and virtually all plants and animals possess adaptations or behaviors for maintaining water balance. Further, the absence of moisture suppresses areal productivity and organic matter decomposition, affects plant distributional patterns, and strongly influences nutrient turnover. Desert life, however, is surprisingly rich and well-adapted to the prevailing climatic conditions. Carnivores and herbivores are generalist feeders, with the latter group sometimes having a significant impact on vegetative composition.

The final section of Chapter 13 describes tundra ecosystems, which are somewhat desert-like with very low precipitation; however, tundras can be distinguished by low temperature (average annual temperature of less than 0°C), which creates a landscape dominated by freezing, thawing and the overall action of frost. These climatic regimes can be found at high altitudes and latitudes, which result in the alpine and arctic tundra, respectively. The harsh climate restricts tundra ecosystems to low-growing, structurally simple plant communities of low productivity and low plant and animal diversity. Plants are dominated by sedges, grasses and shrubs. Tundras also share another common feature with deserts: nitrogen and phosphorus limitation. Many factors interact to restrict pools of available nutrients, e.g., short growing season, low decomposition and low precipitation.

When all is considered, coverage of these five ecosystem types will require at least two or three one-hour lecture periods. The opportunities for emphasizing the close coupling between prevailing climatic regimes and ecosystem type and function are innumerable. Interesting discussions also can be developed around themes relating to the adaptation of plants and animals to their particular environmental setting. Contrasts and parallels can be drawn between the adaptations of organisms between different ecosystem types. Although the information may seem merely descriptive to a seasoned teacher, students often are excited to learn about systems that they may visit one day or have seen already.

TOPICS FOR DISCUSSION

1. Why are fence rows now the primary habitat for birds, invertebrates and small mammals in agricultural regions of the midwestern U.S.?

2. What are some analogies that can be drawn between grazing pressure, plant competition and nutrient availability in natural grasslands and the clipping/maintenance of lawns by humans? Humans remove grass biomass, nutrients and water with mowing, only to have to augment those losses with fertilization and water sprinklers. Plant competition is suppressed by the use of herbicides.

3. Discussions can be developed around facets of coevolution in grasslands, e.g., between herbivores and grasses, between animal prey and predators.

4. What are the ramifications of human encroachment into ecosystems adapted for periodic fire, e.g., grasslands and shrublands? Have students noticed reports of incompatibilities in the newspapers, particularly fire reports from southern California and the western U.S. in late summer? In terms of the economic costs associated with protecting dwellings from fire and the damage to natural systems because of fire suppression, what are opinions regarding home construction in fire-prone habitats?

5. Why are some grasslands suitable for cultivation? What are the environmental consequences of

converting vast tracks of grassland into agricultural production?

6. Contrast the animal adaptations needed for survival in a desert with those needed in a tundra. Why are many animals in grasslands, shrublands and deserts drab in coloration? Why isn't hibernation a common phenomenon among tundra animals?

7. Why are nutrients often limiting in desert, tundra and chaparral ecosystems? How do these systems differ from grasslands? What role does the accumulation of organic matter play in the nutrient cycling of each ecosystem?

8. Using a map of world biogeographic provinces, examine the mosaic of terrestrial communities. Comment on the climatic variations that have created this mosaic. In what regions has man had the most or least impact on the environment? Why?

LECTURE OUTLINE

In addition to providing an opportunity to describe the important ecological features of a variety of common ecosystems, Chapters 13 - 16 also contain information that will allow instructors and students to consider humankind's usage, exploitation and, in many cases, devastation of ecosystems. The extent to which such environmental issues can be explored depends on whether or not environmental issues are an important component of the course being taught. In the outline below, human effects are listed with each ecosystem type; however, an alternative organizational scheme might be to consider environmental degradation of ecosystems in a collective fashion after students have become familiar with the characteristics of a variety of systems.

Many possibilities exist for organizing and combining information between Chapters 13 and 14. The text has subdivided terrestrial ecosystems into forested and unforested systems, but instructors need not feel constrained to this approach. For example, an individual who primarily focuses on ecosystems of the U.S. may find an organizational scheme that progresses along latitudinal and longitudinal gradients more useful. The outline below parallels the text in dealing with unforested ecosystems separately; however, the organizational flow is different. While the text effectively considers the classification, structure and function for each ecosystem separately, the outline below represents an alternative scheme in which structural/functional considerations are discussed after an overview to the various ecosystems is presented; this approach can facilitate comparisons among ecosystem types.

I. Overview to Nonforested Ecosystems

   A. Grasslands
      1. Defining Characteristics
      2. Grassland Types
         a. Prairies
            i.   tall-grass
            ii.   short-grass
            iii. mixed grass
            iv.   palouse
         b. Desert Grassland
         c. Annual Grassland
         d. Successional Grassland
   B. Tropical Savannas

LECTURE OUTLINE CONTINUED:

    C. Shrublands
        1. Defining Characteristics
        2. Shrubland Types
            a. Mediterranean-Type Shrublands
            b. Desert Shrublands
            c. Heath
            d. Successional Shrublands
    D. Deserts
        1. Defining Characteristics
        2. North American Deserts (not covered in text)
    E. Tundra
        1. Defining Characteristics
        2. Tundra Types
            a. Arctic Tundra
            b. Alpine Tundra

II. Structural Comparisons: Grassland, Savanna, Shrubland, Desert, Tundra

    A. Vertical Structure
    B. Root:Shoot Ratios
    C. Plant Distribution
    D. Presence - absence of Decaying Organic Matter
    E. Plant Diversity
    F. Characteristic Animals

III. Functional Comparisons: Grassland, Savanna, Shrubland, Desert, Tundra

    A. Primary Productivity
        1. Relationship to Temperature and Moisture
        2. Ecosystem Comparisons
    B. Nutrient Cycling

IV. Human Impact: Grassland, Savanna, Shrubland, Desert, Tundra

RESOURCE MATERIALS

Important works are cited throughout the chapter. Numerous monographs, however, exist for each ecosystem type and some of these are listed below. With regard to North American ecosystems, refer to Barbour and Billings (1988) for descriptions of ecosystems covered in both Chapters 13 and 14; the emphases in this book are plant species composition, architecture and the overall environment, with separate chapters dedicated to tundras, deserts, shrublands/steppes and forests and mesotropical vegetation. *Introduction to World Vegetation* (Collinson, 1988) also provides a synthesis of world ecosystems, emphasizing ecological principles and organismal adaptations to particular environments. Similarly, Abber and Melillo (1991) provide a good synopsis of terrestrial ecosystems and their structure and function. Another useful source of information on specific ecosystem types can be found in the series entitled *Ecosystems of the World*, published by Elsevier Science Publishing Co., Inc. For the ecosystems described in Chapter 13, see the following volumes in this series: Volume 5 - *Temperate Deserts and Semi-Deserts*; Volume 8A - *Natural Grassland*; Volume 9A & B - *Heathlands and Related Shrublands*; Volume 11 - *Mediterranean-Type Shrublands*; Volume 12A - *Hot Desert and Arid Shrubland*; Volume 13 - *Tropical Savannahs*. For older books that describe the original conditions of grasslands and deserts in North America, refer to Weaver (1954) and Jaeger (1957). In addition,

Bliss et al. (1973) provide information about the ecology (descriptions of abiotic and biotic components) and function of tundras with a review of human impacts. Stanton (1988) provides an interesting perspective on below-ground ecosystems in grasslands with a consideration of the role of roots as resources for these ecosystems.

Aber, J.D. and J.M. Melillo. 1991. *Terrestrial Ecosystems*. Saunders College Publishing, New York.

Bliss, L.C., G.M. Courtin, D.L. Pattie, R.R. Riewe, D.W.A. Whitfield, and P. Widden. 1973. "Arctic tundra ecosystems." *Annual Review of Ecology and Systematics* 4:359-399.

Barbour, M.G. and W.D. Billings. 1988. *North American Terrestrial Vegetation*. Cambridge University Press, New York.

Collinson, A.S. 1988. *Introduction to World Vegetation*. Chapman and Hall, New York.

Jaeger, E.C. 1954. *The North American Deserts*. Stanford University Press, Stanford, CA.

Stanton, N.L. 1988. "The underground in grasslands." *Annual Review of Ecology and Systematics* 19:573-589.

Weaver, J.A. 1954. *The North American Prairie*. The University of Nebraska Press, Lincoln, NB.

Multimedia Resources

Insight Media, 2162 Broadway, New York, NY 10024. --- **"Desert Ecology"** --- 20-minute video.

JLM Visuals, 1208 Bridge Street, Grafton, WI 53204. --- numerous slide sets are available regarding specific ecosystems, e.g., **"Ecology of World Savannas," "Ecology of North American Grasslands," "Ecology of the Everglades,"** etc.

Educational Images Ltd., P.O. Box 3456, West Side, Elmira, NY 14905 --- numerous slides sets are available for specific ecosystems, e.g., **"Ecology of the Arctic Tundra," "Ecology of the Alpine Tundra," "Ecology of the Chaparral," Ecology of the Sonoran Desert,"** etc.

CHAPTER 14

FORESTS: BOREAL TO TROPICAL

CHAPTER SUMMARY AND ORGANIZATION

As its title implies, this chapter describes a general pattern of forest types along a latitudinal basis. The basic pattern is that the extremes between 60° N and equatorial regions are dominated by coniferous forests and tropical forests, respectively, with temperate forest between. Keep in mind, however, that great variation exists, forest types intergrade together in a mosaic and many different types of each forest exist. The text describes the general characteristics and varieties of each of these three forests with each description ending with discussions of forest structure and function. The sections covering forest function detail the processing of energy and nutrients; in some organizational schemes for a general ecology course, an instructor might choose to incorporate some of this information into lectures regarding nutrient cycles/biogeochemistry as specific examples of nutrient pathways from specific ecosystems.

Students should become aware that, because of the dominance of arboreal species, forests illustrate the greatest complexity of terrestrial communities. As this chapter describes, forests are defined and characterized on the basis of species-specific differences in lifeform and structure, i.e., physiognomy (see Chapter 28). In addition to emphasizing the unique ecological characteristics of these systems, instructors also should take the opportunity to emphasize the economic importance of these ecosystems to humankind and the magnitude of exploitation to which these systems have been subjected, e.g., deforestation in the Great Lakes states and the Pacific Northwest, the deforestation of tropical rainforests and various tropical islands, and the list continues.

The first forest type described is the coniferous forest, which is further subdivided into five distinct ecosystems: the taiga, temperate rainforest, montane forest, woodlands and southern pine savannas. By and large, students can be made aware that the existence of coniferous ecosystems can be related typically to regions of modest environmental rigor, either cool temperatures, low rainfall, soils of poor quality and/or fire frequency. Thus, many forests are found at high latitudes or altitudes. Students must take these generalizations with some caution; for example, the temperate rainforest is characterized by high rainfall and mild temperature regimes, yet the soils are thin and poor in nutrients. Similarly, the southeastern pine forests occur in regions of ample moisture and warm temperatures, yet owe their existence to nutrient-poor, sandy soils and fire prevalence.

The largest and most expansive of conifer forests is the taiga (boreal forest), which generally is limited by the summer and winter positions of the Arctic front. The taiga can be subdivided further into four distinct habitat types, two of which represent ecotonal regions of boreal forest-tundra habitat and mixed hardwood-conifer forest along the northerly and southerly limits, respectively. The remaining two habitats are open boreal woodland, dominated by black spruce in damp areas, and the main boreal forest, characterized by continuous stands of white spruce and balsam fir.

The temperate rainforest (sometimes referred to as the Pacific North Coast Forest) is the most luxuriant of the coniferous forest, featuring trees of unparalleled size and longevity (relative to other coniferous forests), e.g., overstory trees can be more than 50 meters tall and have a diameter exceeding two meters; furthermore, longevity is generally greater than 500 years, with some species living more than 1000 years. These exceptional characteristics are clearly related to the excessive moisture regimes (more than 600 cm, with about a third of this amount occurring as fog drip) and mild temperatures.

The montane coniferous forests, of course, are associated with mountainous regions. Instructors may want to divide the coniferous forests of mountainous regions into the subalpine forest at the higher elevations, dominated by spruce and fir species, and the montane forest at the lower elevations, which are dominated by pines (particularly ponderosa pine in the U.S.) and some spruces. Relatively speaking, montane forests are warmer, wetter, more species-rich, and more productive than subalpine forests. Below the montane zone, in areas of western North America that are too dry and hot for montane vegetation, open stands of pinion pine and juniper occur (woodlands).

The southern pine forest (sometimes referred to as the southern pine savannas) occurs in the southeastern United States. Students should understand that these forests are not representative of climax vegetation. Rather, their occurrence on sandy coastal plains and the frequency of fires slow succession and help maintain these communities.

The textbook characterizes three classes of coniferous forest structure: (1) "pines with straight, cylindrical trunks," (2) "spire-shaped evergreens" (spruce and fir) and (3) "pyramid, open crowns," characterized by deciduous conifers (cypress and larch). The physiognomy of these forests is often simple, with stratification typically limited to a canopy layer and a poorly developed forest floor layer of mostly ferns, mosses, dwarf shrubs and a few herbaceous species. A third layer is represented by the often thickly developed litter layer. Because of temperature and moisture limitation, and lower pH regimes, decay is slow and thick litter mats develop over time.

The final section regarding coniferous forests describes the function of these ecosystems in terms of various processes of nutrient accumulation and regeneration. In general, coniferous forests occur on soils of poor nutrient quality, and decomposition of organic matter is slow. For these reasons, nutrients tend to accumulate in conifers, and in some cases, the nutrients are scavenged directly from rainfall via algae and lichens, some of the latter of which are capable of nitrogen fixation. Most plant decay is carried out by fungi, which are stimulated by vernal pollen deposition. The text draws comparisons between nitrogen cycling in young and old Douglas fir stands of the Pacific Northwest and between Douglas fir stands and loblolly pine communities of the southeastern U.S. coastal plains. In both the mature and immature fir stands, input of nitrogen is similar and quite low, but the compartmentalization of nitrogen is very different. Relative to the immature stand, the old growth forest accumulates more nitrogen with larger pools of nitrogen occurring in root and bole. On the other hand, the soil of old growth forests contains a smaller percentage of the total nitrogen pool and a smaller proportion of the total detrital organic matter (because of accumulation of slower decomposing plant materials in the older forests). In contrast to the fir forests, loblolly pine grows more quickly and accumulates more nitrogen in root tissues. Nitrogen accumulated in plant biomass represents 11%, 8% and 13% of the total nitrogen pool in the young fir forest, old fir forest and loblolly forest, respectively. In other words, a greater proportion of available nutrients are sequestered at an early age in loblolly pine. Data facilitating these comparisons are presented in Tables 14.1 - 14.3; incorporation of this material into lectures will facilitate comparisons as well as help students visualize the biogeochemical processes occurring in these systems.

The temperate broadleaf forests occur in three general forms: temperate deciduous forests, temperate woodlands and temperate evergreen forests. Students should understand that these forests have been used extensively by humankind for centuries; hence, most of what they may see today are second- and third-growth forests. In light of the winter and summer climatic extremes, which must be endured by species in these forests, application of the term "temperate" is somewhat misleading.

The temperate deciduous forest once covered large sections of several continents. While these forests are still common, European and Asian forests have been the most impacted by clearing for various purposes. Among the continents, the European forests are the most species depauperate because of glacial patterns. In North America, the deciduous forests can be divided into nine regional associations (Barbour et al., 1987 - see **Resource Materials**); the most prevalent of these is described briefly in the textbook. Two noteworthy associations are the mixed mesophytic forest and the oak-chestnut forest. The mixed mesophytic forest is noted for its species richness, which can exceed 25 species per hectare. The oak-chestnut forest is a dramatic example of the actions of humankind on ecosystems. The chestnut was virtually eliminated by the introduction of chestnut blight (*Endothia parasitica*) during the early part of the 20th century.

The temperate woodlands are represented by oak, oak-sagebrush or oak-pine communities in the southwestern U.S. and Mexico. By comparison, temperate evergreen forests occur in subtropical areas, e.g., the Florida keys, Australia, South America, etc. Typical species include magnolias, palms, eucalyptus, live oaks, and other tree species.

Mature, uneven-aged, temperate forests are clearly stratified. Typically, four layers are visible (canopy, subcanopy, shrub layer and ground layer). These strata are less well-defined in disturbed even-aged stands. Such zonation influences gradients of light, temperature and humidity. In summer, the canopy is well-lighted, warm and typically less humid, in contrast to the forest floor, which is shaded, humid and somewhat cooler. Of course, the patterns are strongly related to the seasonal patterns of leaf presence/absence and spatial patterns of closed and open forest (gaps). These zones can be traversed freely by some animals, such as birds; yet, other animals of lesser mobility (particularly insects) sometimes are confined to one or two layers. In any event, the greatest faunal diversity occurs on or in the ground layer.

Several points are emphasized in the concluding section regarding the functioning of temperate forests. First, the uptake of nutrients by trees rarely meets requirements and must be augmented by internal recycling within tree biomass. Second, the most important nutrient pool is the litter layer, which has a turnover time of about four years. Third, these ecosystems tend to accumulate large amounts of nitrogen, most of which is in the upper horizons of the soil layer. Differences are drawn between the rates of organic matter decomposition of oak-pine forests and yellow poplar-oak forests; rates of carbon degradation are about twice as fast in the yellow poplar system. Again, students should be asked to draw comparisons between the data in Tables 14.5 - 14.7. The discussion of forest function ends with a comparison of temperate broadleaf forests and evergreen forests. The key difference between these systems is that nutrients are retained and cycled more slowly in coniferous trees. Undoubtedly, this relates to the evergreen nature of these trees and is the primary reason why they often are described as nutrient accumulators.

The average student will have virtually no familiarity with the composition, heterogeneity or extent of tropical forests; most students, however, probably will have some awareness of tropical deforestation. Because of this absence of familiarity, the use of visual materials, such as slides, films, etc., will prove beneficial to helping students relate to the information in the textbook and lectures. Students should understand that not all tropical forests are rainforests. Forests of the tropics come in many varieties and the textbook sequentially describes tropical rainforests, seasonal forests, and dry forests. Examples of the structural and functional aspects of tropical forests, however, are primarily for the rainforests.

With regard to rainforests, points to be stressed are: (1) the exceptional plant and animal diversity, (2) the magnitude of net production relative to temperate forests, (3) the vegetative stratification patterns (often five layers), (4) the gradients of microclimatic factors, e.g., CO_2, temperature, light, humidity, (5) the variety of lifeforms, e.g., lianas and epiphytes, (6) the importance of disturbance in successional patterns and maintenance of species, and (7) the complexity of interactions that can exist among plant and animal species. Lecture material regarding the complexity of interactions and the high diversity can be coupled with information regarding deforestation to help students fully understand the effects of indiscriminate destruction of these forests.

From a functional standpoint, the energetics of tropical rainforests are such that more than 70% of the gross production is used in maintenance. The plant species also are supported by very rapid turnover of organic material and nutrients. In some cases, mycorrhizal fungi form links between decaying organic matter and living plant roots. Not all tropical forests, however, occur in nutrient-poor soils, where such adaptations are necessary.

TOPICS FOR DISCUSSION

1. Discuss the wisdom (or lack thereof) of clear-cutting (see Chapter 29 for an overview of tree harvesting strategies). In particular, ask students to consider the ecological implications associated with a harvesting practice that creates forests of trees of homogeneous age and size. Of course, other points also can be raised, such as loss of habitat, biomass, nutrients and species diversity.

2. Why are forests valuable to humankind's existence? For each forest type, what are the products that man uses; don't necessarily confine the discussion to "products" that have conventional economic value. Do forest processes have inherent value to life on this planet, e.g., CO_2 removal?

3. On a broad level, some animal species inhabit the edges of forests while other species occur in the deep interior of forests. How have urbanization, agriculture and forestry probably affected the distributions and abundance of organisms in each of these groups? A good example in the northeastern United States is the whitetail deer (an edge species). Why are whitetail deer so abundant in many states?

4. Try to stimulate students to think in a landscape perspective. For example, how does the destruction and/or disturbance (natural or human-induced) affect aquatic communities within forests or aquatic communities that receive run-off from forests?

5. What are (have been) some of the cultural consequences of deforestation for aboriginal humans, who have coexisted with forest species (see Bennett, 1992 - **Resource Materials**)?

LECTURE OUTLINE

I. Introduction

    A. Classification of Forests
        1. Differences in Life-Form and Layering
        2. Role of Climate and Soil
    B. Types of Forests
        1. Coniferous Forests
        2. Temperate Broad-leaved Forests
        3. Tropical Forests
    C. Vertical Microclimatic Gradients in Forests

II. Conifer Forests

    A. Types
        1. Taiga
        2. Temperate Rainforest
        3. Montane Forests
        4. Woodlands
        5. Southeastern Pine Forests
    B. Structural Classes
        1. Straight, Cylindrical Trunks
        2. Spire-shaped Evergreens
        3. Deciduous Conifers
    C. Functional Considerations
        1. Biomass Production and Nutrient Usage
            a. Young versus Mature Douglas Fir Stands
            b. Douglas Fir versus Loblolly Pine Stands
        2. Decomposition of Organic Matter
        3. Importance of Woody Litterfall

LECTURE OUTLINE CONTINUED:

III. Temperate Broad-leaved Forests

A. Types
  1. Temperate Deciduous Forests
  2. Temperate Woodlands
  3. Temperate Evergreen Forests
B. Arrangement of Layers
  1. Even-aged Versus Uneven-aged Stands
  2. Distribution of Animal Life
C. Functional Considerations
  1. Internal Recycling
  2. Litter Decay and Nutrient Turnover

IV. Tropical Forests

A. Types
  1. Rainforests
  2. Seasonal Forests
  3. Dry Forests
B. Structural Considerations
  1. Unique Growth-forms
  2. Influence of Layering on Microclimate and Faunal Distribution
  3. Importance of Tree Fall to Structural Diversity
C. Functional Considerations
  1. Comparative Net Productivity (relative to boreal and temperate forests)
  2. Rapidity of Litter Decay

V. Examples and Discussion of Deforestation by Forest Type

RESOURCE MATERIALS:

Packham et al. (1992) offer much information about the interactions between various forest ecosystem components. For consideration of North American forests, refer to Perlin (1991). Also related to North American forests (temperate forests) is the classic work describing studies in the Hubbard Brook Experimental forest (Bormann and Likens, 1979). With regard to tropical rain forests, the Chapman and Hall company has published three books about rainforest ecology and the politics of exploitation and deforestation (Mabberley, 1991; Park, 1992; Whitmore and Sayer, 1992). Further information about nutrient cycling in moist tropical forests can be found in the work of Vitousek and Sanford (1986). Murphy and Lugo (1986) review the ecology of dry tropical forests. Bonan and Shugard (1989) offer a review paper covering many environmental factors and their effect on boreal forests.

Many of the monographs about terrestrial ecosystems provided for Chapter 13 are also relevant to Chapter 14, e.g., Barbour and Billings (1988), Collinson (1988) and Abber and Melillo (1991). In addition, *Ecosystems of the World*, published by Elsevier Science Publishing Co., Inc., provides some volumes about forest biomes: Volume 10 - *Temperate Broad-Leaved Evergreen Forests*, and Volumes 14 A & B - *Tropical Rain Forest Ecosystems*.

Aber, J.D. and J.M. Melillo. 1991. *Terrestrial Ecosystems*. Saunders College Publishing, New York.

Barbour, M.G., J.H. Burk, and W.D. Pitts. 1987. *Terrestrial Plant Ecology* (2nd Edition). Benjamin/Cummings Publishing Company, Menlo Park, CA.

Barbour, M.G. and W.D. Billings. 1988. *North American Terrestrial Vegetation*. Cambridge University Press, New York.

Bennett, B.C. 1992. "Plants and people of the Amazonian rainforests." *Bioscience* 42:599-607.

Bonan, G.B. and H.H. Shugart. 1989. "Environmental factors and ecological processes in boreal forests." *Annual Review of Ecology and Systematics* 20:1-28.

Bormann, F.H. and G.E. Likens. 1979. *Pattern and Process in a Forested Ecosystem*. Springer-Verlag Inc., New York.

Collinson, A.S. 1988. *Introduction to World Vegetation*. Chapman and Hall, New York.

Mabberley, D.J. 1991. *Tropical Rain Forest Ecology* (2nd Edition). Chapman and Hall, New York.

Murphy, P.G. and A.E. Lugo. 1986. "Ecology of tropical dry forest." *Annual Review of Ecology and Systematics* 17:67-88.

Perlin, J. 1991. *Forest Journey*. Harvard University Press, Cambridge, MA.

Packham, J.R., D.J.L. Harding, G.M. Hilton, and R.A. Studdard. 1992. *Functional Ecology of Woodlands and Forests*. Chapman and Hall, New York.

Park, C.C. 1992. *Tropical Rainforests*. Chapman and Hall, New York.

Van Cleve, K., F.S. Chapin, C.T. Dyrness, and L.A. Viereck. 1991. "Element cycling in taiga forests: state-factor control." *Bioscience* 41:78-88.

Vitousek, P.M. and R.L. Sanford. 1986. "Nutrient cycling in moist tropical forest." *Annual Review of Ecology and Systematics* 17:137-167.

Whitmore, T.C. and J.A. Sayer. 1992. *Tropical Deforestation and Species Extinction*. Chapman and Hall, New York.

Multimedia Resources

Biology Media, 2700 York Road, Burlington, N.C. 27215. --- **"The Moist Coniferous Forest"** --- 35 mm. slide set.

Insight Media, 2162 Broadway, New York, NY 10024. --- **"Coniferous Forest"** --- 20-minute video.

JLM Visuals, 1208 Bridge Street, Grafton, WI 53204. --- numerous slide sets are available regarding specific ecosystems, e.g., **"Ecology of the Deciduous Forest," "Ecology of Pine Barrens,"** etc.

Educational Images Ltd., P.O. Box 3456, West Side, Elmira, NY 14905 --- numerous slides sets available for specific ecosystems, e.g., **"The Tropical Rainforest," "Tropical Rain Forests Under Fire,"** Ecology of the Northeast Forest," etc.

CHAPTER 15

FRESHWATER ECOSYSTEMS

CHAPTER SUMMARY AND ORGANIZATION

The organizational scheme used in Chapters 13 and 14 for examining terrestrial communities is extended to aquatic systems in Chapter 15, which covers freshwater communities, and Chapter 16, which describes marine ecosystems. The importance of a thorough understanding of aquatic systems cannot be overstated. In a social, cultural and environmental context, the study of, and our knowledge of, freshwater processes ultimately will be vital to maintaining the quality of human life. Water is fundamental to all life on Earth, but mismanagement of this essential resource threatens nearly every region of the planet. Ultimately, human survival and maintenance of a reasonable standard of living will require the efficient use and management of our water resources (Wetzel, 1992- see **Resource Materials**). Recognition of this problem today allows us to prepare students with the knowledge needed to make educated decisions in the future about this valuable resource.

Chapter 15 maintains the pattern established in Chapters 13 and 14, whereby an initial overview of each ecosystem type is provided and structural and functional aspects subsequently are considered. In general, two types of aquatic ecosystems exist, those without appreciable flow (lentic systems) and those with flow (lotic systems). The former group includes lakes, ponds and wetlands; wetlands are treated separately by the text because of their unique dominance by hydrophytic vegetation and a hydrology that is markedly different from lakes. Systems with flow include streams and rivers; reservoirs, as dammed lotic systems, also are included in the section of flowing aquatic systems.

The chapter initially characterizes the different mechanisms by which lakes are created. Lake formation is followed by discussion of temperature patterns, oxygen, CO_2 (bound and unbound) and light. In a very real way, temperature patterns, which are affected by light penetration and wind mixing, affect the vertical zonation of lakes (i.e., stratification), which in turn affects patterns of oxygen, carbon dioxide and other nutrients. Hence, instructors may find it easiest to discuss some of these characteristics in a broader context of physical structure. Furthermore, ideas about nutrient processes and production can be integrated into this discussion.

Because most students may perceive lakes as homogenous bodies of water, they must be introduced to the fact that lakes are characterized by horizontal and vertical gradients, some of which can be used to characterize various patterns of zonation. Vertical gradients of light, temperature and oxygen should be described with emphasis on how these influence the distribution of organisms, particularly the phytoplankton and zooplankton. If students have already been introduced to the concept of trophic status (eutrophic versus oligotrophic), a useful approach is to contrast differences in vertical gradients between the two types of lakes, particularly nutrient patterns and oxygen profiles. In this context, the manner in which primary production and decomposition affect the distribution of oxygen and some nutrients can be explained. In addition, the seasonal changes in some vertical profiles can be discussed in the context of lake turnover. No matter what organizational scheme is used, productive, nutrient-rich, eutrophic lakes must be contrasted to their nutrient-poor, oligotrophic counterparts, which are characterized by low productivity.

Horizontal zonation is most noticeable in the littoral zone of lakes where subdivisions are based on vegetation type: emergent plants, floating plants and submersed plants. These zones have a pronounced effect on lake ecosystems by providing habitats for numerous invertebrate species and by functioning as a source of organic carbon (dissolved and particulate) and nutrients to the open water zone (limnetic zone).

Students should become aware that lakes are not isolated systems; they are profoundly influenced by inputs from outside the lake basin via wind-borne deposition, groundwater seepage, surface runoff, and precipitation. Internally derived materials are described as "autochthonous" while external inputs are termed "allochthonous." In the normal functioning of lakes, the important allochthonous inputs for lake metabolism are nutrients and organic carbon, but anthropogenic pollutants can have a major impact on water bodies. For instance, the primary source of mercury to the Great Lakes comes from airborn particles resulting from the combustion of fossils fuels in the midwestern U.S.

Many important aspects of lotic systems exist that can be contrasted to the lentic environment; five key characteristics should be covered in lectures. Obviously, the first aspect is that the degree of flow influences all aspects of the ecosystem, including organismal type and characteristics, nutrient cycling, channel characteristics, etc. Second, the heterotrophic nature of lotic systems needs to be discussed. The contribution of autotrophs to energy processing in these systems is negligible because water flow inhibits phytoplankton and scours periphyton from areas of fast flow; furthermore, deposition of silt inhibits periphyton development in pools where flow is slow (students should be introduced to the differences that exist between riffles and pools). Third, students should be introduced to the various functional groups of animals in streams and rivers and how their abundance varies along the reach of a lotic system from headwaters to a river's mouth. The latter idea can be used to introduce the river continuum concept, which characterizes the gradual changes that occur over a lotic system's reach; in addition to organismal changes, differences in flow rate, temperature, and magnitude of primary productivity can be discussed in the continuum concept. Fourth, students should understand stream order and, fifth, lectures should include a description of nutrient spiraling as an additional unique aspect of lotic systems.

The textbook concludes the section on lentic systems by considering "regulated" rivers and streams. In particular, the impacts of damming and reservoir creation are characterized. Reservoirs have characteristics of both lentic and lotic systems. For example, constant inflow into the lake and sometimes flow throughout the reach of the reservoir (depending on operation of the dam) is reminiscent of a lotic system; yet, the flow is usually insufficient to prevent at least the occasional occurrence of temperature stratification and its concomitant effects. Although many positive benefits to humankind can be derived from reservoirs, the biological impacts can be dramatic. Impoundment interferes with nutrient cycling and transport (reservoirs trap nutrients and rapidly become eutrophic). Water flow downstream of a reservoir is often irregular, characterized by pulses of high flow, which scours the river channel, and periods of little or no flow, which allows the channel to dry. Furthermore, the quality of water released from a reservoir is rarely suited to the requirements of lentic species. These are only a few of the impacts of dams; many other impacts relate to the drowning of terrestrial communities.

Among nonscientists, wetland ecosystems are probably one of the most under-appreciated, least understood and most fiercely debated habitats. Debate about wetlands focuses primarily on the legal definition of a wetland that is used in protective legislation and in litigation. The need for strong protective legislation in the face of overwhelming pressure to drain wetlands is paramount; 53% of wetlands already have been drained in the continental U.S., with loss rates currently estimated at about 450,000 acres/year. Wetlands must be characterized as transitional habitats, somewhere between terrestrial and aquatic habitats. Transitional habitats are particularly difficult to characterize and almost impossible to define precisely. Nonetheless, as the textbook describes, wetlands are described on the basis of hydrology, hydrophytic vegetation and hydric soils. The text also describes why this three-pronged approach is necessary and the problem with relying entirely on vegetative characteristics. Clearly, the most important attribute is the hydrology. Variations in these three characteristics can be used to classify wetlands into various types. The text describes differences between numerous wetland types, such as basin wetlands, riverine wetlands, fringe wetlands, marshes, swamps, mires, riparian wetlands and bogs.

Most structural features of a wetland can be related to hydrology; hydrology includes both movement of water and hydroperiod. Hydroperiod, in particular, influences vegetation type, zones and cycles, as well as affecting rates of organic matter production and decomposition. Nutrient processing in wetlands can be exceedingly complex, varying between wetland types and varying as a function of numerous factors, such as the nature of the watershed, vegetation and soils, the magnitude of groundwater input versus surface flow and precipitation, and the overall climatic regime. The textbook describes nutrient processing in two wetland types: a freshwater marsh and a peatland. Nutrient processing within a freshwater marsh is closely related to the rates of vegetative production and decay; in general, wetland plants act as nutrient pumps moving elements from the substratum to surface waters. Peatlands, by contrast, are much less productive and turn nutrients over at a very slow rate. These characteristics are related to the significant storage of nutrients in undecomposed organic matter (peat) and the extraordinarily slow decay processes (95% turnover time of peat can be about 3,000 years).

Given today's legislative context of wetland regulation, no lecture about wetlands would be complete without stressing the need for wetland protection. Rather than focusing on just the historical losses of wetlands, instructors need to try to instill a sense of wetland appreciation in their students. This can be approached by emphasizing the many positive aspects of wetlands; the most dramatic contributions of wetlands include water purification, groundwater recharge, storage of flood waters, waterfowl production, the provision of habitats for numerous endangered species, fishery spawning grounds, and food chain support for various aquatic and terrestrial species.

TOPICS FOR DISCUSSION

1. What are some physical and chemical properties of lotic systems that are affected by the clear-cutting of forests along channel banks and in drainage basins? Try to encourage students to think about management of aquatic ecosystems in a landscape perspective.

2. In what way does the human language portray and influence man's thinking about the quality of wetlands? Some wetlands are etymologically unique among ecosystems in being both nouns and verbs, e.g., bogs, mires and swamps. How does such usage color human perception of these ecosystems?

3. How is the biota of lakes influenced by the stocking of lakes, ponds and rivers with carnivorous game fish?

4. What is the value of placing buffer zones of undisturbed vegetation around the edges of lakes, ponds and wetlands?

5. How does oxygen availability affect the cycling of nitrogen and sulfur in lake ecosystems? Contrast processes between eutrophic and oligotrophic lakes. Why is aeration often used in the management of artificial ponds and some reservoirs?

6. Is the river continuum concept affected by the creation of dammed reservoirs? If so, in what ways? What influence do reservoirs have on nutrient spiraling?

LECTURE OUTLINE

I. Initial Definitions

    A. Limnology
    B. Lentic
    C. Lotic

II. Lentic Systems

    A. Selected Organismal Groups
        1. Phytoplankton
        2. Zooplankton
        3. Benthos
    B. Vertical Gradients and Horizontal Zonation
        1. Profile of Light Penetration
            a. Photic Depth
            b. Tropholytic versus Trophogenic Zones
        2. Profile of Temperature Change
            a. Epiliminion, Metalimnion, and Hypolimnion
            b. Thermocline and Depth of Wind Mixing
        3. Profile of Oxygen Change
        4. Horizontal Zonation in the Littoral Zone
    C. Selected Elemental Considerations
        1. Relationship between CO_2, alkalinity and pH
        2. Nutrient/Trophic Status
            a. Eutrophic
            b. Oligotrophic
            c. Dystrophic
    D. Seasonal Overturn
        1. Breakdown of Vertical Gradients
        2. Redistribution of Nutrients

III. Lotic Systems

    A. Importance of Flow: Pools versus Riffles
    B. Stream Orders
    C. Processing of Organic Carbon
        1. Heterotrophy and the Processing of CPOM and FPOM
        2. Functional Groups of Organisms
            a. Shredders
            b. Collectors
            c. Scrapers
            d. Piercers
            e. Decomposers
    D. Nutrient Spiraling
    E. The River Continuum Concept
    F. Reservoirs: An Example of River Regulation

IV. Wetlands

    A. Defining Characteristics
        1. Hydrology
        2. Hydrophytic Vegetation
        3. Hydric Soils

LECTURE OUTLINE CONTINUED:

B. Wetland Types
 1. Basin Wetlands, Riverine Wetlands and Fringe Wetlands
 2. Marshes, Swamps and Bogs
 3. Mires: Fens, Bogs and Moors
C. Energy and Nutrient Processing
 1. Freshwater Marsh
 2. Peatland
D. Discussion of Wetland Protection
 1. Patterns of Wetland Loss
 2. Wetland Value: Positive Functions

RESOURCE MATERIALS

The literature on aquatic ecosystems is vast, an indication of the importance of these systems. The four primary works used in limnology courses are by Wetzel (1983), Barnes and Mann (1991), Cole (1994) and Horne and Goldman (1994). Of these, Wetzel's book is the most comprehensive reference work, but does not describe lotic systems. The remaining three books, however, each have chapters about lotic systems. Brief overviews to the functioning of stream communities can also be found in the works of Cummins (1974) and Minshall (1980); additional information regarding rivers and streams can be obtained from the classic work by Hynes (1970), Barnes and Minshall (1983) and Townsend (1980). Harper and Ferguson (1995) offer a collection of river management papers; topics include flow manipulations, water quality considerations and management of vegetation, fish and mammals. In addition, volume 45, issue 3 of *Bioscience* is dedicated to the ecology of large rivers.

The limnology textbooks cited above are the best sources for detailed summaries of the structure and function of lakes and ponds. The definitive reference work in limnology which should not be overlooked is the four volume *Treatise on Limnology* by G.E. Hutchinson. These monographs provide detailed information and a wealth of references on virtually all aspects of limnology. One of the few references to reservoir limnology is Thornton et al. (1990).

Information on wetlands is finally becoming consolidated. The best overall reference regarding wetlands is the textbook by Mitsch and Gosselink (1994). A more general introduction to wetland ecosystems can be found in the book by Weller (1981). More specific technical symposium proceedings can be found in literature by Good et al. (1978), Greeson et al. (1978), Sharitz and Gibbons (1989,) van der Valk (1989) and Mitsch (1994). In addition, the Society of Wetland Scientists has their own journal, entitled *Wetlands*, which covers the gamut from applied to theoretical works on wetland ecology.

The problems of water degradation are placed into clear perspective by Wetzel (1992). Multiple facets of aquatic contamination are described in a comprehensive textbook by Laws (1993). Additionally, Hynes (1974) serves as a reference to pollution problems primarily in rivers. A comprehensive survey of the effects of nitrogen loading in wetlands is also addressed by Morris (1991).

Lotic Systems

Barnes, J.R. and G.W. Minshall (eds.). 1983. *Stream Ecology: Application and Testing of General Ecological Theory*. Plenum Press, New York.

Cummins, K.W. 1974. "Structure and function of stream ecosystems." *Bioscience* 24:631-641.

Fontaine, T.D. and S.M. Bartell. 1983. *Dynamics of Lotic Ecosystems*. Ann Arbor Science, Ann Arbor, MI.

Hynes, H.B.N. 1970. *The Ecology of Running Waters*. University of Toronto Press, Toronto.

Minshall, G.W. 1980. "Autotrophy in stream ecosystems." *Bioscience* 28:767-771.

Townsend, C.R. 1980. *The Ecology Stream and Rivers*. Edward Arnold, London.

Lentic Systems

Barnes, R.S.K. and K.H. Mann. 1991. *Fundamentals of Aquatic Ecology*. Blackwell Scientific Publication, London.

Cole, G.A. 1994. *Textbook of Limnology* (4th edition). Waveland Press, Inc., Prospect Heights, Illinois.

Horne, A.J. and C.R. Goldman. 1994. *Limnology* (2nd Edition). McGraw-Hill, Inc., New York.

Hutchinson, G.E. 1975-1994. *A Treatise on Limnology*: Volume I - *Geography, Physics and Chemistry*; Volume II - *Introduction Lake Biology and the Limnoplankton*; Volume III - *Limnological Botany*; Volume IV - *The Zoobenthos*. John Wiley and Sons, Inc., New York.

Thornton, K.W., B.L. Kimmel, and F.E. Payne. 1990. *Reservoir Limnology: Ecological Perspectives*. John Wiley and Sons, Inc., New York.

Wetzel, R.G. 1984. *Limnology* (2nd edition). Saunders College Publishing, New York.

Wetlands

Greeson, P.E., J.R. Clark, and J.E. Clark (eds.) 1978. *Wetland Functions and Values: The State of our Understanding*. American Water Resources Association, Minneapolis, MN.

Harper, D.M. and A.J.D. Ferguson. 1995. *The Ecological Basis for River Management*. John Wiley and Sons, Inc., New York.

Larsen, J.A. 1982. *Ecology of the Northern Lowland Bogs and Conifer Forests*. Academic Press, New York.

Mitsch, W.J. (Ed.) 1994. *Global Wetlands: Old World and New*. Elsevier Publishing, Amsterdam.

Mitsch, W.J. and J.G. Gosselink. 1994. *Wetlands* (2nd Edition). Van Nostrand Reinhold Co., New York.

Sharitz, R.R. and J.W. Gibbons (eds.). 1989. *Freshwater Wetlands and Wildlife*. U.S. Department of Energy (available as DE90005384 from the National Technical Information Service, Springfield, VA).

Weller, M.W. 1981. *Freshwater Marshes - Ecology and Wildlife Management*. University of Minnesota Press, Minneapolis.

van der Valk, A. (Ed.). 1989. *Northern Prairie Wetlands*. Iowa State University Press, Ames, IA.

Aquatic Pollution

Hynes, H.B.N. 1974. *The Biology of Polluted Waters.* University of Toronto Press, Toronto.

Laws, E.A. 1993. *Aquatic Pollution - An Introductory Text* (2nd Edition). John Wiley and Sons, Inc., New York.

Morris, J.T. 1991. "Effects of nitrogen loading on wetland ecosystems with particular reference to atmospheric deposition." *Annual Review of Ecology and Systematics* 22:257-279.

Wetzel, R.G. 1992. "Clean water: a fading resource." *Hydrobiologia* 243/244: 21-30.

Multimedia Resources

Insight Media, 2162 Broadway, New York, NY 10024. --- videos available, e.g., **Lakes and Streams**" and "**Bog Ecology.**"

JLM Visuals, 1208 Bridge Street, Grafton, WI 53204. --- various 35 mm slide sets available, e.g., "**Survey of Freshwater Communities**," "**Life of a Pond**," "**Pond Life**," "**Pond Animals**," etc.

Educational Images Ltd., P.O. Box 3456, West Side, Elmira, NY 14905 --- numerous slide sets available, e.g., "**The World of a Lake**," "**The World of a River**," "**Freshwater Biology**," "**Ecology of a Stream**," "**Bog Ecology**," etc.

CHAPTER 16

MARINE ECOSYSTEMS

CHAPTER SUMMARY AND ORGANIZATION

As the final chapter dealing with a survey of ecosystems types, Chapter 16 represents a brief overview of seven marine systems. Because 70% of the Earth's surface is covered by oceans, one chapter cannot do justice to the diversity, complexity and importance of these systems. This chapter is important because it provides students with a basic understanding that oceans are not homogenous systems; biotic communities just as structurally distinct and often more productive and diverse than terrestrial ecosystems occur beneath the seemingly barren surface waters and along shorelines. Students also should understand that the size of oceans does not make them invulnerable to degradation from human activity. In fact, just the opposite is occurring as ocean basins often function as the ultimate depository of many human wastes and pollutants carried in by rivers and the atmosphere; in some cases, wastes have been deliberately dumped in ocean waters. Furthermore, many marine species have suffered from overharvesting and unwise management, offshore drilling, nuclear testing, and other offenses.

Chapter 16 begins with a description of the various vertical and horizontal strata of marine habitats. Various terms are introduced, such as the oceanic provinces, the neretic, oceanic, abyssal, and hadapelagic; this brief introduction sets the stage for understanding that oceans obviously are not well-mixed, uniform systems. The opening introduction also details other unique characteristics of marine habitats to which marine organisms are superbly adapted; namely, high salinity, variations in pressure, and unique hydrodynamic features of water movement. Students should become aware that these conditions are far different from a terrestrial existence (*e.g.*, how many terrestrial organisms have to be able to exist or tolerate multiple atmospheres of pressure?). After the introduction of these important physical features, seven different marine habitats are described including the open ocean, coral reefs and various coastal systems. Each section is organized on the basis of the structural and functional attributes of the ecosystem.

Phytoplankton, zooplankton, nekton and benthic communities are described as the main components of biotic structure in the open ocean. Although each of these groups also occurs in freshwater habitats, there are interesting differences, such as the presence of bioluminescent algae and fish, extraordinary examples of vertical migration by zooplankton (hundreds of meters in only a few hours), the occurrence of toxic red tides, and amazing benthic communities clustered around hydrothermal vents. Two key functional attributes that must be emphasized are (1) the overall low oceanic productivity except in specific regions, particularly upwelling areas, and (2) the importance of the microbial loops, in which food chains are supported by bacterial utilization of dissolved organic carbon released by phytoplankton. Students should realize that the microbial loop is a significant variation from energy processing in terrestrial habitats. The text also provides an overview to variations in net productivity and the composition of oceanic food chains.

The primary structural feature of rocky shorelines is the striking zonation, which is characterized by particular animals. These shores are characterized by three basic zones and their subdivisions: supralittoral fringe, littoral zone, and infralittoral fringe. Students should be able to identify and recognize the zones on the basis of animal life. Furthermore, students should develop an appreciation of the dramatic hydrological regimes that dominate rocky shores with the perpetual washing by tides and waves. Exposed at low tide, tide pools filled with organisms experience dramatic fluctuations in salinity, temperature and oxygen.

Structure in mudflats and sandy shores is not readily obvious since most organisms spend a great proportion (if not all) of their existence beneath the mud/sand. Emphasize differences between this subterranean refuge and exposure to fluctuating water levels and environmental gradients along rocky shores; for example, in terms of temperature and salinity, a more stable habitat exists a few centimeters into the substrate than can be found on rocky shorelines. Zones along sandy and muddy shores (supralittoral, littoral and infralittoral) are differentiated by the meiofauna. Functionally, the critical factor for life in these habitats is the accumulation of organic matter. Food chains are strongly heterotrophic and are supported by the bacterial production associated with organic matter decomposition; primary productivity is low and confined to the intertidal zone. The text emphasizes that these systems should be considered as part of the entire coastal ecosystem encompassing salt marshes, estuaries and coastal waters.

Coral reefs are unique, richly diverse communities found in warm, shallow waters along continental shelves in subtropical to tropical waters. The complexity and diversity of these communities are too vast to be covered in detail in any general textbook, however, several ideas need emphasis. Corals are both heterotrophic and autotrophic, involving a symbiotic relationship between the coral and endozoic algae; much work has been done on the reciprocal effect of these organisms on one another. Diversity and productivity gradients in coral habitats are defined by light penetration, but it is interesting to note that productivity along coral reefs can be a hundredfold greater than in adjacent oceanic waters.

Estuaries are unique systems in which freshwater rivers empty into the sea; hence, the structural aspects of estuaries are shaped by currents and salinity. As the textbook infers, about the only generalization that can be made about estuaries is that they are dynamic systems in which the physical environment exhibits extreme spatial and temporal variation. Logically, salinity is lowest at the river's entrance to an estuary. However, it varies both vertically and horizontally, primarily as a function of currents which are themselves influenced by rainfall, wind and tides. Furthermore, temperature patterns show daily and seasonal fluctuations. For organisms, survival depends on osmotic adjustment to the salinity variations and an ability to cope with currents. In fact, the distribution of many organisms is limited by salinity gradients. Energetic and nutrient pathways are supported by plankton production, detrital production and internal nutrient cycling. Rooted aquatic plants also assume importance in some shallow estuaries. Although all marine ecosystems have been negatively impacted by anthropogenic activity, estuaries have been particularly vulnerable (see Kennish, 1992 - **Resource Materials**); rivers are all too often sources of human-generated contaminants.

The structure of salt marshes is partitioned into aquatic and semi-aquatic zones. Plant dominance is determined by salinity gradients, tidal effects, freshwater intrusions, and their interactions. In North American wetlands, these systems can be divided into the low marsh in the intertidal region and the high marsh, which is less influenced by tide and where higher temperatures, evaporative potential and salinity prevail. The production of salt marshes is extremely high (as much as 2800 g dry weight/m^2/yr) with the grazing food chain dominated by the ribbed mussel, fiddler crab, periwinkle and smaller invertebrates, birds and snails. Most of the primary production (about 90%), however, is processed by the detrital food chain, in which microbial oxidation of organic matter occurs primarily in anaerobic soils. Students will need to rely on information learned in Chapter 12 to understand that oxidative processes under anaerobic conditions can be supported by the reduction of sulfate and nitrate. The reduction of nitrate to nitrogen gas (denitrification) is the primary reason that most salt marshes function as nitrogen sinks. The textbook provides a detailed example of nitrogen cycling for a salt marsh. Phosphorus supplies are less critical in salt marshes and generally exceed requirements.

The final shoreline ecosystem described in Chapter 16 is the mangrove wetland. Mangroves are confined to subtropical and tropical shores where wave action is minimal. Pronounced zonation in these systems is determined by the extent of tidal flooding and surface runoff. Examples of zonation are provided for both subtropical and tropical mangroves, with zonation patterns being more pronounced in the tropical systems. Similar to salt marshes, net productivity is high in mangrove ecosystems and the dynamics of energy and nutrient processing are under the influence of tidal washing, water and soil nutrients, and salinity. Detrital exports from mangroves to adjacent estuaries provide an energetic base for fish and invertebrate species.

TOPICS FOR DISCUSSION

1. Why are some marine habitats, such as coral reefs, salt marshes and mangroves, so phenomenally productive while the vast portion of the open ocean is poorly productive?

2. How has the construction of dams on freshwater rivers near seas affected populations of anadromous fish? How can agricultural activity far removed from an ocean ultimately degrade marine systems? Emphasize that the biological effects of altering one ecosystem are seldom localized. Furthermore, emphasize interactions between terrestrial and aquatic ecosystems.

3. What has happened to many coastal marshes and mangrove systems? Ask students to consider the fact that many cities are built at the mouth of estuaries. What kinds of systems were destroyed for development of these population centers? What are student perceptions about the mentality that such ecological systems are wastelands with little net value?

4. Some species use beaches for nesting purposes; two prominent examples are the piping plover and the pacific green sea turtle. Why are these two species endangered and what role has human utilization of sandy shores played? What are the students' attitudes about human access to all publicly-owned beach sites?

5. Are tropical coral reefs in some ways the marine equivalent to tropical rainforests? What are some ways in which this analogy might hold true, e.g., diversity?...productivity?...highly coevolved species?...influence of light on zonation? What are some important ways, however, that this analogy might prove false (e.g., role of disturbance in maintaining species diversity)?

LECTURE OUTLINE

I. Vertical and Horizontal Zonation Oceans

A. Open Ocean
B. Benthic Region

II. Physical Influences on Oceanic Life

A. Salinity
B. Pressure
C. Water Movement
    1. Waves
    2. Langmuir Currents
    3. Upwelling
    4. Tides

III. Survey of Ecosystem Types: Structure and Function

- A. Open Ocean
    - 1. Biotic Structure
        - a. Phytoplankton
        - b. Zooplankton
        - c. Nekton
        - d. Benthos
            - *i.* diversity and variation
            - *ii.* hydrothermal vents
    - 2. Functional Attributes
        - a. Magnitude of Productivity
        - b. Regions of Upwelling
        - c. Microbial Loop
- B. Rocky Shores
    - 1. Horizontal Zonations
        - a. Supralittoral Zone
        - b. Littoral Zone
        - c. Infralittoral Zone
        - d. Sublittoral Zone
    - 2. Tide Pools
    - 3. Functional Attributes
        - a. Productivity
        - b. Significance of Wave Action Disturbance
- C. Sandy Shorelines and Mudflat Communities
    - 1. Biotic Structure
        - a. Epifauna
        - b. Infauna/meiofauna
        - c. Horizontal Gradients: Supralittoral, Littoral and Infralittoral
    - 2. Organic Matter Accumulation and Ecosystem Function
- D. Coral Reefs
    - 1. Types
        - a. Fringing Reefs
        - b. Barrier Reefs
        - c. Atolls
    - 2. Biotic Structure
        - a. Corals and Zooanthellae
        - b. Light Gradients
    - 3. Autotrophic and Heterotrophic Nature of Coral Function
- E. Estuaries
    - 1. Structural Complexity and Influence on Biota
        - a. Variations in Salinity
        - b. Current, Tides and Mixing Patterns
    - 2. Estuarine Food Chains and Function
- F. Salt Marshes
    - 1. Biotic Structure
        - a. Low Marsh versus High Marsh
        - b. Salt Pans
    - 2. Salt Marsh Function
        - a. Magnitude of Productivity: Tidal Subsidy
        - b. Sulfate and Nitrate Reduction/Oxidation of Organic Matter
        - c. Role as a Nitrogen Sink

LECTURE OUTLINE CONTINUED:

  G. Mangrove Wetlands
    1. Horizontal Zonation and Mangrove Development
    2. Functional Attributes
      a. Magnitude of Productivity
      b. Influence of Hydroperiod
      c. Export of Organic Matter
      d. Commercial Value

IV. Examples of Environmental Degradation

  A. Oil Contamination and Spills
  B. Overharvesting of Marine Species
  C. Loss of Habitat (marshes, nesting grounds, coral reefs)
  D. Potential Influence of Global Warming
  E. Contaminant-laden Rivers
  F. Garbage and Sewage

RESOURCE MATERIALS

Instructors will not have to look far to discover a wealth of literature about marine ecosystems. Jumars (1993) provides an excellent introduction to numerous biological/ecological facets of oceanography. In addition, Ross (1988) provides an excellent beginning textbook in oceanography. For a perspective about recent changes in marine biodiversity and human causes, see *Understanding Marine Biodiversity* (1995). Volume 41, issue 7 of *Bioscience* is also dedicated to a treatment of marine biological diversity. A classic, popular work with an environmental perspective is Rachel Carson's work, *The Sea Around Us* (reprinted 1989). Technical review articles can be found in the *Annual Review of Oceanography and Marine Biology* (Hafner Press) and *Advances in Marine Biology* (Academic Press). More general articles can be found in the periodical, *Oceans*.

The collection of titles below about estuarine/salt marsh ecology range from technical to general. Chapman (1977) and Ketchum (1983) are part of Elsevier's series on ecosystems of the world. McLusky (1971) offers a useful description of estuarine plants and animals and the estuarine environment. Perkins (1974) also provides good descriptions of the abiotic and biotic components of estuaries. Pomeroy and Wiegert (1981) divide their material into three sections dealing with the structure and function of salt marshes, salt marsh populations and the salt marsh ecosystem (cycles, models, diversity, etc.). Kennedy (1980) provides a collection of papers, some of which are technical, dealing with estuarine chemical cycles, primary production and ecosystem dynamics. For a good overview to the various ways that humans have impacted estuarine ecosystems, consult Kennish (1992). Technical articles can be found in the journal, *Estuaries*.

The strength of the material by Stephenson and Stephenson (1972) lies in their description of specific intertidal communities on the planet. Similarly, Mathieson et al. (1991) provide information about specific locations as well as a general introduction to intertidal communities.

Information about coral reefs is less consolidated. However, chapters 20 and 21 in the book by Levinton (1982) describe the biology of coral reefs. Falkowski (1993) describes effects of nutrient enrichment on coral symbiosis. Ladd (1961) and Mark (1976) provide short introductions to reef building processes.

Oceanography - Marine Biology

Carson, R. 1989. *The Sea Around Us*. Oxford University Press, Oxford.

Committee on Biological Diversity in Marine Systems. 1995. *Understanding Marine Biodiversity*. National Academy Press, Washington D.C.

Jumars, P.A. 1993. *Concepts in Biological Oceanography: An Interdisciplinary Primer*. Oxford University Press, Oxford.

Ross, D.A. 1988. *Introduction to Oceanography* (4th Edition). Prentice Hall, Inc., Englewood Cliffs, NJ.

Estuaries - Salt Marshes

Chapman, V.J. (editor). 1977. *Wet Coastal Ecosystems*. Elsevier Science Publishers, Amsterdam, The Netherlands.

Kennedy, V.S. (editor). 1980. *Estuarine Perspectives*. Academic Press, New York.

Kennish, M.J. 1992. *Ecology of Estuaries: Anthropogenic Effects*. CRC Press, Ann Arbor, MI.

Ketchum, B.H. (editor). 1983. *Estuaries and Enclosed Seas*. Elsevier Science Publishers, Amsterdam, The Netherlands.

McLusky, D.S. 1971. *Ecology of Estuaries*. Heinemann Educational Books Ltd., London.

Perkins, E.J. 1974. *The Biology of Estuaries and Coastal Waters*. Academic Press, New York.

Pomeroy, L.R. and R.G. Wiegert. 1981. *The Ecology of a Salt Marsh*. Springer-Verlag, New York.

Intertidal Zones

Mathieson, A.C. and P.H. Nienhuis (editors). 1991. *Intertidal and Littoral Ecosystems*. Elsevier Science Publishers, Amsterdam, The Netherlands.

Stephenson, T.A. and A. Stephenson. 1972. *Life Between the Tidemarks on Rocky Shores*. W.H. Freeman & Co., San Francisco, CA.

Coral Reefs

Buddemeier, R.W. and D.G. Fautin. 1993. "Coral Bleaching as an Adaptive Mechanism." *Bioscience* 43:320-326.

Falkowski, P.G., Z. Dubinsky, L. Muscatine, and L. McCloskey. 1993. "Population control in symbiotic corals." *Bioscience* 43:606-611.

Mark, K. 1976. "Coral Reefs, Seamounts, and Guyots." *Sea Frontiers* 22:143-149.

Ladd, H.S. 1961. "Reef Building." *Science* 15:703-715.

Mangroves

Jerome, L.E. 1977. "Trees that help build the land." *Oceans* 10:39-45.

Teas, H.J. (editor). 1984. *Physiology and Management of Mangroves*. DR W. Junk Publishers, Dordrecht, The Netherlands.

Tomlinson, P.B. 1986. *The Botany of Mangroves*. Cambridge University Press, Cambridge.

Multimedia Resources

Insight Media, 2162 Broadway, New York, NY 10024. --- videos available, e.g., **Secrets of the Salt Marsh**" and "**Estuaries.**"

JLM Visuals, 1208 Bridge Street, Grafton, WI 53204. --- various 35 mm slide sets available, e.g., "**The Salt Water Marsh**," and "**Life of the Oceans**."

Educational Images Ltd., P.O. Box 3456, West Side, Elmira, NY 14905 --- numerous slides sets available, e.g., "**The World Ocean**," "**Land Beneath the Sea**," "**Marine Productivity, Food Webs and Nutrient Cycles**," etc.

CHAPTER 17

PROPERTIES OF POPULATIONS

CHAPTER SUMMARY AND ORGANIZATION

Chapter 17 establishes a foundation for understanding many aspects of population ecology, a topic which dominates the following 10 chapters of the textbook. In particular, information in Chapter 17 will be relevant to later considerations of population growth (Chapter 18), life history patterns (Chapter 20) and population genetics (Chapter 21). As the different properties of populations are discussed in the classroom, you should emphasize that each of these characteristics has biological reality; in other words, population characteristics tell us something about a species: how it functions in its environment and how it interacts with members of its own species and other species. For example, a simple measure such as density can provide a basis for speculating (certainly not definitively) about a species' position in a food chain, the life history strategy of that species, and possible stress on a population as perhaps related to items such as food availability or disease. As another example, survivorship curves impart fundamental information about the life history strategy of an organism which can then be related to competitive ability, longevity, parental care, etc. Use the ideas and terminology in this chapter as a way of learning about the functioning of various organisms, but, as always, emphasize that interpretations must be exercised with caution.

The chapter begins with an introduction to basic ideas about what constitutes a population. Use the terminology described to emphasize the functional aspects of populations. In particular, stress the importance of recognizing populations as both genetic and ecological units in which species' members interact with one another. In fact, it can be argued that a population's characteristics are very significantly influenced by the positive and negative interactions of its members among one another and with members of other species. The significance of interactions can be carried over into a classroom discussion of the second section of Chapter 17, which covers density and dispersion. Clumped and uniform dispersion patterns certainly can be used to stimulate a conversation about potential interactions among individuals and/or interactions between individuals and their environment. Clumping might only result from poor dispersal capabilities, but in many examples, aggregations of individuals exist because of higher social behavior or a patchy distribution of resources. Uniform distributions of individuals almost always arise from interactions (or at least the desire to suppress interactions) in homogenous environments, e.g., territoriality or allelopathy. Given that species interact with the biotic and abiotic components of their environment, it shouldn't come as any surprise that random distributions are the most rare. In general, a random distribution indicates the absence of any patterns of attraction or repulsion with respect to where an organism finds itself in its habitat. As these distribution patterns are discussed, contrast the ideas of fine-grained and coarse-grained distributions. In addition, emphasize that spatial distributions also can have a temporal facet to their existence, which may be related to when offspring disperse or may vary as a function of migratory behavior.

The final sections of Chapter 17 cover age structure, sex ratios, mortality and natality, respectively. The relationship between age structure and mortality and natality should be presented in the classroom. Obviously, a stable age distribution is dependent on the existence of equal age-specific birth and death rates (an unlikely occurrence). Age structure distributions sometimes are used as one tool in wildlife management for predicting future changes in a population. This sort of usage, however, is dependent on mortality and natality schedules remaining the same over extended periods of time (another unlikely occurrence).

A major section of the textbook's discussion of mortality and natality focuses on life tables. Some instructors will not feel it necessary for their students to be able to construct a life table. Life tables, however, do have useful applications in an introductory ecology course. First, a portion of any ecology course should deal with population growth and regulation (Chapter 18 in the text). One life table statistic, the net reproductive rate (R_o), provides a first indication of population growth from the point of view of whether or not an individual (usually a female) is replacing itself through reproduction during its life. Life tables also provide the basis for constructing population projection tables (life history tables), which can be used to model population growth (assuming stable survivorship and fecundity). Second, a life table provides valuable information for understanding some aspects of how an organism is functioning in its environment. For instance, the life table can be the basis for constructing and interpreting survivorship and mortality curves and for building fecundity tables from which the reproductive value of different age groups can be estimated. As Chapter 17 illustrates, lectures should include some discussion of how survivorship curves seldom fit one of the ideal theoretical curves.

Throughout population ecology, it is important to emphasize some fundamental differences between the population biology of plants and animals. Chapter 17 provides several opportunities to make comparisons. Examples of unitary versus modular populations will involve, in part, comparisons between some plant and animal systems. Furthermore, the discussions of age structure and natality in the textbook have specific sections that focus on the unique aspects of plant species as they differ from animals.

TOPICS FOR DISCUSSION

1. Is the spatial scale at which a population dispersion pattern is evaluated important in interpretation of the pattern? Clams might be distributed randomly in a mudbank, but are mudbanks randomly distributed over the planet's surface?

2. Use a local cemetery to construct a life history table. What can be surmised about age- specific mortality and natality on the basis of this table? How do birth, death and longevity compare today with populations of 80 or more years ago?

3. Use some acceptable technique to quantify population density in optimal and marginal habitats for an invertebrate species. Mark-recapture techniques might be suitable. Is density an indication of the quality of the habitat? Can these observations be related to environmental reasons for the existence of clumped distributions?

4. Consider perennial plants versus annual plants and the ways in which each plant type differs from various animal species. Contrast their age structure patterns, and their mortality and natality.

LECTURE OUTLINE

I. Introduction

    A. Terms and Definitions
       1. Demography
       2. Modular Versus Unitary Populations
       3. Ramets Versus Genets

LECTURE OUTLINE CONTINUED:

    B. Populations as Genetic Units
    C. Populations as Demes

II. Population Properties

    A. Density
        1. Crude Density
        2. Ecological Density
    B. Dispersion Pattern
        1. Random
        2. Clumped
        3. Uniform
        4. Fine-grained Versus Coarse-grained Patterns
        5. Spatial versus Temporal Dispersion Patterns
    C. Age Structure
        1. Stable Versus Stationary Age Distributions
        2. Differences in Age Structure Between Plants and Animals
    D. Sex Ratios: Changes Among Age Groups
    E. Mortality and Natality
        1. Terminology
            a. Death Rate
            b. Probability of Living Versus Probability of Dying
            c. Physiological Versus Realized Natality
            d. Crude versus Specific Birthrates
            e. Age-specific Schedule of Births
        2. Life Tables
            a. Types
                i.   dynamic
                ii. dynamic-composite
                iii. time-specific
            b. Components
                i. l_x - known survival
                ii. q_x - proportional mortality
                iii. m_x - fecundity
                iv. R_o - net reproductive rate
            c. Survivorship Curves
                i.   type I
                ii. type II
                iii. type III
            d. Mortality Curves
            e. Fecundity Curves
            f. Comparison of Life Tables Between Plants and Animals

RESOURCE MATERIALS

Hedrick (1984) covers virtually all facets of population ecology, including separate chapters about population growth, population genetics, life history phenomena, interspecific competition, predation, etc. Instructors also will find information in the book by Pianka (1994) applicable to most topics of Chapter 17.

The works by Murray (1979) and Boughey (1973) are older references that offer introductions to classic population ecology; most of the topics in these books are relevant to a general ecology course. Silverton (1993) has written a recent work on plant population biology. Vandermeer

(1981) includes chapters that describe life tables; however, a minimal working knowledge of calculus and matrix algebra is needed.

Collections of classic works can be found in the book edited by Hazen (1975); particularly relevant to Chapter 17 is Deevey's paper regarding animal life tables. As a collection of some of the most important papers in ecology, Real and Brown (1991) include particularly noteworthy population papers with commentary.

Boughey, A.S. 1973. *Ecology of Populations* (2nd Edition). The Macmillan Company, New York.

Hazen, W.E. (editor). 1975. *Readings in Population and Community Ecology*. W.B. Saunders Company, Philadelphia.

Hedrick, P.W. 1984. *Population Biology - The Evolution and Ecology of Populations*. Jones and Bartlett Publishers, Inc., Boston.

Murray, B.G. 1979. *Population Dynamics: Alternative Models*. Academic Press, New York.

Pianka, E.R. 1994. *Evolutionary Ecology* (4th Edition). HarperCollins College Publishers, New York.

Real, L.A. and J.H. Brown. 1991. *Foundations of Ecology*. University of Chicago Press, Chicago, IL.

Silverton, J.W. 1993. *Introduction to Plant Population Biology*. Blackwell Scientific Publications, Boston.

Vandermeer, J. 1981. *Elementary Mathematical Ecology*. John Wiley and Sons. New York.

Multimedia Resources

PDRS Computer Systems, Box 301, Glassboro, NJ, 08028. --- **"Dispersion Analysis"** --- allows assessment of how organisms are dispersed.

CHAPTER 18

POPULATION GROWTH AND REGULATION

CHAPTER SUMMARY AND ORGANIZATION

The information in Chapter 18 provides both a strong basis for understanding various aspects of theoretical ecology and understanding some of the consequences of overpopulation, particularly in the context of human ecology. The first two sections of the chapter explain mathematical formulations, which either describe the direction of population growth or project a simulation/model of population growth. Virtually all of the information in the first section regarding "rate of increase" will require a thorough familiarity with life tables as presented in Chapter 17. Specifically, information from a life table provides the basis for constructing a population projection table, an age distribution table, and r, which represents the intrinsic rate of natural increase. Furthermore, the population projection table is needed to calculate λ, usually termed the finite rate of increase, which then can be used to predict geometric population growth. The math is logical and based on elementary algebra, but a preliminary understanding of life tables will be needed.

The information described in the first section provides three indices, which serve as indicators of the direction of population growth. R_o indicates whether an organism is replacing itself; hence, an $R_o = 1$ indicates a stable population, while values greater than or less than 1 represent growing and declining populations, respectively. The symbol λ represents a ratio of population size between successive time periods. Again, a ratio of 1 shows no population growth (i.e., stable population), while values greater than and less than 1 are indicative of growing and declining populations, respectively. The intrinsic rate of natural increase, r, represents differences between instantaneous rates of birth and death. If birth rate = death rate, then $r = 0$. This means the net change in population size is zero and the population is stable. When $r > 0$, a population is growing, and if $r < 0$ the population is declining. Of course, there are assumptions implicit in using each of these indices, but the idea is that these represent predictors of population growth.

The second section of the chapter covering population growth encompasses a presentation of the exponential and logistic growth models. Both models use r as a component, but the logistic model differs in that it includes the carrying capacity of the environment (K), which ultimately limits population size. The limitations and assumptions inherent in each of these models should be discussed in the classroom. Emphasize that the exponential models are only applicable to initial growth after colonization of an unexploited habitat by a species; hence, exponential growth is a transient phenomenon. Also, students should understand that K is not a fixed value; rather, it represents a theoretical expression of an equilibrium population size around which population density fluctuates. Lecture discussions about population size as it relates to fluctuations around K necessarily will include a consideration of time lags.

Although too simplistic, the logistic growth model is particularly valuable as a standard by which to compare actual changes in population density, and an example is provided. The model, however, is also valuable from an instructional viewpoint because it can be used to provide a "systems" context for understanding population regulation. In other words, the ideas of positive and negative feedback in systems (introduced in Chapter 3) now can be applied to the logistic growth model and can be used to stimulate a discussion/lecture of population regulation.

Many details of density-dependent population regulation in the context of feedback mechanisms are described in the third section of Chapter 18. Keep in mind that population regulation is not the same as population limitation. Density-dependent regulation affects growth in proportion to population size. The fundamental process in population regulation is intraspecific competition. Multiple facets of intraspecific competition are discussed with presentations of the types of competition, its effects on growth and fecundity (emphasizing animals), and various aspects of competition among members of plant species. Some instructors may want to incorporate a portion or all of this information in their presentation of Chapter 19, which encompasses a detailed presentation of other aspects of intraspecific competition. On the other hand, this information serves as a good introduction of topics to come.

The discussion of density-dependent regulation is followed with a section about density-independent influence on population size. Be sure that students recognize that, although density-independent factors can certainly affect population growth, they certainly *do not regulate* population growth. Regulation involves homeostatic feedback. Nonetheless, population sizes are affected by the interaction of density-dependent and density-independent processes. Weather patterns are used as an example of an important density-independent factor. In some instances, density-independent factors can be more important than density-dependent regulation in determining population size. For example, how important is density-dependent regulation when predation holds a population below a density at which intraspecific competition is important?

The fifth and sixth sections of the chapter introduce new ideas relevant to understanding population change and regulation. First, population "fluctuations" in comparison to "cycles" are characterized; the chief difference between these two terms is essentially a function of the regularity at which population maxima and minima occur. The magnitude of fluctuations in population density vary with a species' resilience to disturbance, which often is related to reproductive rates. Fluctuations are usually localized events; cyclic fluctuations are typically confined to structurally and functionally simple ecosystems. Second, key factor analysis is described as a means of determining the specific causes of density-dependent mortality; thus, it would be appropriate to include key factor analysis in the discussion of density-dependent regulation.

The final section of the chapter provides an overview to extinction. Students should understand that although the actions of humans have resulted in mass extinctions worldwide, extinction is a natural process that has always occurred, even in epochs preceding human rise. The extinction of a species usually occurs as the accumulation of local population extinctions (i.e., regional mass extinctions are rare). Small, isolated populations endemic to specific habitats are more vulnerable to extinction. The textbook contrasts deterministic and stochastic extinctions as well as providing information about some causes and examples of extinction. Although the models of population growth basically force students to think about growing populations, students should realize that, ultimately, at some time, all species become extinct. Extinction is not a desirable occurrence from a human perspective, but a biological reality nonetheless.

TOPICS FOR DISCUSSION

1. How has technology relieved human populations somewhat from normal constraints on population growth? Can advances in agricultural technology keep up with the nutritional needs of an exponentially-growing human population? What are some recent examples of countries in which local human populations seemingly have been under density-dependent control? What is the probability that the human population in its planetary totality eventually will be subject to density-dependent control? What are some examples of density-independent factors that affect localized human population densities? (For information about human carrying capacity, see Daily and Ehrlich, 1992 - **Resource Materials**.)

2. In examples of biological control, predators often reduce prey to very low levels. Clearly, the density of both populations is an important factor in this interaction, but is density-dependent regulation likely to be important in affecting the prey's population size?

3. In terms of management of game species and protection of endangered species, what is the importance of understanding factors that affect population size?

4. One of the ideas for preserving tropical forests is to create multiple small-scale habitat preserves, essentially habitat islands, instead of setting aside a few expansive tracts of forest. How might the fragmentation of the habitat into small parcels ultimately affect extinction processes? How important to your answer is a consideration of the distance between these preserves?

LECTURE OUTLINE

Some instructors may want to unify the information in the outline below with the detailed information regarding intraspecific competition presented in the following chapter (Chapter 19). No matter whether intraspecific competition is integrated into lectures about population growth or in some other context (e.g., behavioral ecology), this competitive interaction should be emphasized as the primary *regulatory* mechanism governing population density.

I. Models of Population Growth

    A. Geometric Growth (based on λ)
    B. Exponential Growth (based on r)
    C. Logistic Growth and the Carrying Capacity
        1. K
        2. Time-lags

II. Determinants of Population Size

    A. Density Dependent Regulation
        1. Positive and Negative Feedback on Population Growth
        2. Importance of Intraspecific Competition
            a. Types of Competition: Exploitative and Interference
            b. Animal Response to Intraspecific Competition
                i. growth rates
                ii. fecundity
            c. Plant Response to Intraspecific Competition
                i. law of constant yield
                ii. self-thinning: -3/2 power law
        3. Key Factor Analysis
    B. Density-independent Effects on Population Size
        1. Climate
        2. Food Supply

III. Population Fluctuations and Cycles

    A. Causes
        1. Time Lags
        2. Extrinsic Influences
            a. Weather
            b. Food Supply
            c. Interactions With Other Organisms (e.g., parasitism)

LECTURE OUTLINE CONTINUED:

     B. Resilience
     C. Extinctions
         1. Deterministic versus Stochastic Extinction
         2. Historical Perspective: Mass Extinctions
         3. Contemporary Extinction Rate

RESOURCE MATERIALS

Key references are provided throughout Chapter 19. In addition, some of the references presented in the **Resource Materials** of Chapter 18 in this guide are also relevant (e.g., Vandermeer, 1981; Murray, 1979); Murray, in particular, provides an overview to models and density-dependent limits on population growth. Lomnicki (1988) presents a reductionist approach to population ecology with a focus on individuals. Flowerdew (1987) presents information about life tables, key-factor analysis and population dynamics. Tamarin (1978) offers a collection of key papers about the abiotic and biotic regulation of populations; this book also includes information on key factor analysis. Good recent papers dealing with population regulation have been written by Murdoch (1994) and Alcakaya (1992). Westoby (1984) summarizes the history and theory of the -3/2 power thinning rule.

Alcakaya, H.R. 1992. "Population cycles of mammals: Evidence for a ratio-dependent hypothesis." *Ecological Monographs* 62:119-142.

Daily, G.C. and P.R. Ehrlich. 1992. "Population, Sustainability, and Earth's Carrying Capacity." *Bioscience* 42:761-771.

Flowerdew, J.R. 1987. *Mammals - Their Reproductive Biology and Population Ecology*. Edward Arnold, London.

Lomnicki, A. 1988. *Population Ecology of Individuals*. Princeton University Press, NJ.

Murdoch, W.M. 1994. "Population regulation in theory and practice." *Ecology* 75:271-287.

Tamarin, R.H. (editor). 1978. *Population Regulation*. Dowdin, Hutchinson and Ross, Inc., Stroudsburg, PA.

Westoby, M. 1984. "The self-thinning rule." *Advances in Ecological Research* 14:167-225.

Multimedia Resources

Exeter Software, 100 North Country Road, Setauket, New York, 11733. --- **"RAMAS/age"** --- a computer model of population fluctuations in age structured populations; **"EcoSim"** --- a computer program that performs a variety of simulations, including geometric population growth, exponential population growth and logistic population growth.

PDRS Computer Systems, Box 301, Glassboro, NJ, 08028. --- **"Population Growth Analysis"** --- allows examination of population growth with manipulation of progeny, number of generations, etc.

CHAPTER 19

INTRASPECIFIC COMPETITION

CHAPTER SUMMARY AND ORGANIZATION

Expanding on Chapter 18's introduction to intraspecific competition, Chapter 19 provides detailed information on the many effects manifested by intraspecific competition. Most of the chapter is dedicated to life history events and to density-driven behavioral patterns in animals, e.g., territoriality and social hierarchies. In addition to descriptions of these intraspecific interactions, a sub-theme of population regulation runs through the chapter and provides further continuity with Chapter 18. The manner in which various behaviors/events do or do not act as population regulation mechanisms is discussed.

The population-regulating effects of social stress is the first topic presented. Lectures should emphasize that stress functions as a regulation mechanism in animal populations, since it involves feedback mechanisms through the endocrine system. Although plants exhibit many responses to stress, no conclusion regarding density-dependent population regulation by stress in plant populations is possible.

The driving force for dispersal in populations is generally competition for any number of resources; consequently, most dispersal occurs at high population density. Nonetheless, dispersal does not function as a population regulation mechanism because the degree of dispersal is not related to rates of change in population density. Dispersal does provide advantages, such as expanding the ranges of organisms, increasing gene flow, and reducing inbreeding. The probability that these benefits will be realized depends on the age and condition of the dispersers. Clearly, the likelihood that a disperser will be of sufficient condition to colonize a new area successfully depends on the timing of dispersal, the distance dispersed and the dispersal age. The textbook discusses each of these topics in detail. In particular, during presaturation dispersal, or dispersal before the carrying capacity is reached, dispersers are often in good condition since resource availability is still adequate. Dispersal after the carrying capacity has been reached or exceeded is termed saturation dispersal, and the dispersing individuals are usually in poorer condition. Smith also provides examples regarding the distances that dispersers may travel.

Social interactions are considered from the standpoint of social dominance and territoriality. Social interactions are complex, but at their simplest, these interactions maintain a social hierarchy that is sustained by the behavioral patterns of dominance and submissiveness exhibited by each individual. Dominant individuals have maximum access to resources and consequently experience higher fitness than subdominants. If social dominance affects patterns of natality and mortality in a density-dependent manner, it then can be considered a form of population regulation. This section concludes with a detailed example of wolf social structure.

Various kinds of territories exist among animals. For example, some species may defend feeding territories while others defend breeding territories. On the other hand, some species establish a territory in which all activities occur. Territoriality might act as a means of population regulation, but only when territories have a lower size limit. In addition to considering territoriality as a potential means of regulation, classroom discussions about territoriality are best approached in the context of cost:benefit ratios. Defense of territories is energetically expensive and may involve intimidation displays, vocal or chemical signals, and sometimes physical confrontation. These costs, however, must be balanced against the benefits of occupying the territory. Ultimately, increased reproductive success is the primary benefit of occupying an optimal territory. Obviously, the costs of occupying a territory increase with the territorial size. In some species, territorial size is fixed, while in others it varies as a function of density and overall habitat quality. Occasionally, a substantial pool of individuals do not successfully occupy a territory; these individuals create a surplus of individuals, termed a floating reserve. Individuals from this reserve quickly occupy vacancies if a territory is vacated.

The discussion on territoriality concludes with a description of home ranges. Defined as the area where an animal lives, homes ranges are not to be confused with territoriality and may or may not involve aggressive behavior and dominance. A number of generalizations about home ranges can be discussed in class. For example, home range size is related to body size, habitat quality, carnivory versus herbivory, organismal age, and gender. One of the primary benefits of a home range is the familiarity that an individual can develop with his habitat, particularly with respect to availability of food and shelter.

TOPICS FOR DISCUSSION

1. Do populations benefit from territoriality because the risk of the population overexploiting the habitat is minimized, or do certain individuals benefit from the advantages of increased resource availability, reproduction and survival? From an evolutionary standpoint which is more important? Arguments about the evolution of territoriality on the basis of population versus individual benefits were once intensely debated among ecologists.

2. What are some of the similarities between perils of dispersal for animal species and perils faced by human colonists who came to the New World? What are some similarities in terms of risks and reception of benefits?

3. Although a few studies of dispersal exist, their numbers are relatively lower compared to investigations of other aspects of animal behavior. Why? What makes dispersal particularly difficult to study?

4. Students usually think of dispersal involving active means of transport. What are some examples of both plants and animals that use passive means of dispersal (dispersal by water and air movement). Would you describe the dispersal that results from these mechanisms as random? Why not?

5. In some animal groups, males are the primary dispersers, while in other animal groups females disperse. What mating systems would contribute to male dispersal versus female dispersal (see Chapter 20 of textbook for further background information about mating systems)?

LECTURE OUTLINE

I. Dispersal

    A. Background
           1. Natal Dispersal
           2. Breeding Dispersal
           3. Dispersal Distance
           4. Population Expansion versus Population Regulation

LECTURE OUTLINE CONTINUED:

    B. Primary Advantages
        1. Reduced Overcrowding and Social Stress
        2. Reduced Competition with Relatives
        3. Reduced Inbreeding
        4. Enhanced Fitness
    C. Primary Disadvantages
        1. Loss of Familiar Social Environment
        2. Loss of Familiar Physical Environment
        3. Outbreeding Depression
        4. Lower Fitness: Alleles Poorly Adapted to New Environment
        5. Increased Vulnerability During Dispersal

II. Sociality

    A. Social Dominance
        1. Peck Orders
        2. Dominance as a Population Regulator
            a. Influence on Reproduction/Survival
            b. Example: Wolf Packs
    B. Territoriality
        1. Use of Territory
            a. General Purpose
            b. Breeding
            c. Nesting
            d. Feeding
        2. Territorial Defense
            a. Mechanisms
                i.   vocalizations
                ii. confrontation
                iii. chemical
            b. Benefits
                i.   resource acquisition
                ii. mate attraction
                iii. increased fitness: survival and reproduction
            c. Territorial Size
                i.   fixed size
                ii. Huxley elastic disk model
                iii. cost:benefit ratios and optimal size
        3. Floaters
        4. Effect on Population Regulation
    C. Home Range
        1. Influence of Body Size on Range Size
        2. Advantages of a Home Range

RESOURCE MATERIALS

Alcock (1994) offers a good general textbook about animal behavior that presents a concise overview to territorial and homing behavior. Studies of territoriality in insects, birds and mammals are abundant. For examples of insect studies, Baker (1983) provides a good review. Some studies of territoriality in specific invertebrate taxa include works by Whitham (1987, aphid), Lederhouse (1982, butterfly), Fincke (1992, damselfly), and Greenfield et al. (1987, grasshopper). For studies of territoriality in birds, see Gass et al. (1976, hummingbird) and Smith and Shugart (1987, ovenbird). Osterfield (1986) provides an analysis of territoriality in voles.

For a collection of papers regarding dispersal refer to Chepko-Sade and Halpen (1987). Bergstrom (1988), Swihart et al. (1988) and Spencer and Cameron (1990) provide technical studies of mammalian home ranges.

Alcock, J. 1994. *Animal Behavior* (5th Edition). Sinauer Associates, Inc., Sunderland, MA.

Baker, R.R. 1983. " Insect Territoriality." *Annual Review of Entomology* 28:65-89.

Bergstrom, B.J. 1988. "Home ranges of three species of chipmunks (*Tamias*) as assessed by radiotelemetry and grid trapping." *Journal of Mammalogy* 69:190-193.

Chepko, B.B. and Z.T. Halpen (editors). 1987. *Mammalian Dispersal Patterns*. University of Chicago Press, Chicago, IL.

Fincke, O.M. 1992. "Consequences of larval ecology for territoriality and reproductive success of a neotropical damselfly." *Ecology* 73:449-462.

Gas, C.L., C. Angeher, and J. Centra. 1976. "Regulation of food supply by feeding territoriality in the Rufous Hummingbird." *Canadian Journal of Zoology* 54:2046-2054.

Greenfield, M.D. and T.E. Shelly. 1987. "Variation in host-plant quality: implications for territoriality in a desert grasshopper." *Ecology* 68:828-838.

Lederhouse, R.C. 1982. "Territorial defense and lek behavior of the black swallowtail butterfly, *Papiolio polyxenes*." *Behavioral Ecology and Sociobiology* 10:109-118.

Ostfeld, R.S. 1986. "Territoriality and mating system of California voles." *Journal of Animal Ecology* 56:691-706.

Smith, T.M. and H.H. Shugart. 1987. "Territory size variation in the ovenbird: the role of habitat structure." *Ecology* 68:695-704.

Spencer, S.R. and G.N. Cameron. 1990. "Operationally defining home range: temporal dependence exhibited by hispid cotton rats." *Ecology* 71:1817-1822.

Swihart, R.K., N.A. Slade, and B.J. Bergstrom. 1988. "Relating body size to the rate of home range use in mammals." *Ecology* 69:393-399.

Whitham, T.G. 1986. "Costs and benefits of territoriality: Behavioral and reproductive release by competing aphids." *Ecology* 67:139-147.

CHAPTER 20

LIFE HISTORY PATTERNS

CHAPTER SUMMARY AND ORGANIZATION

The content of Chapter 20 is firmly based in evolutionary ecology; consequently, classroom discussions of this information should focus on how different life history phenomena maximize the fitness of individuals in a species. Although the amount of variation in reproductive systems is incredibly great, the key idea is that for these systems to have arisen and continue, they must be selectively advantageous under some set of conditions. Thus, the primary elements of the chapter deal with mate choice, sexual selection and parental investment in reproduction, all of which affect fitness. The latter topic can be approached from a cost:benefit perspective, the costs represented by all energetic and nutritional expenditures related to mating and producing/rearing offspring, while the benefits, of course, are in the form of more offspring and higher fitness.

Organisms reproduce either by sexual or asexual means. Sexual reproduction is the more expensive reproductive mechanism, but it is essential at some point, even in a primarily asexual species, because of the benefits of genetic recombination resulting in new genotypes. Sexual reproduction comes in a variety of forms; attention should be given to Table 20.1 to ensure that students have a complete understanding of the various sexual mating systems. Mate choice, termed sexual selection, is a key component of many sexual systems. Depending on the species involved, intrasexual selection may occur in which males compete for the privilege of mating with a female (numerous examples are presented for consideration in the classroom); in contrast, intersexual selection involves mate choice between the genders (usually the choice is made by females). How females make their choices is explored in detail by the text. The choice may be based on a male's ability to provide resources or its genetic quality, or perhaps a combination of the two.

When to reproduce, how often to reproduce and how much effort to place into parental care are also key questions in reproductive biology. Some organisms have evolved to reproduce multiple times in their life-times (iteroparity), while other species reproduce only once per lifetime (semelparity). Similarly, some species give birth to precocial offspring (needing little care), while others rear altricial offspring (helpless offspring). Ask students to relate each of these systems to environmental conditions and to the constraints of a species' life history, e.g., of what importance would life-span longevity be to a discussion of reproductive behavior? In these discussions, students should not fail to consider plant reproduction and its various forms.

Energy budgets put real constraints on how much a parent can afford to place into producing offspring. Reproductive costs reduce energy for growth and maintenance of the parents; hence, energetic allocations must be made between these different expenditures. To illustrate energetic allocations, the textbook presents an overview of clutch size in birds and some of the various experimental manipulations that have been performed to determine how clutch size affects parental performance and survivorship of offspring. Topics for emphasis relate to the basic choice of how many offspring can be afforded, brood reduction, siblicide and various factors to which clutch size may be related, such as food supply, seasonal variation in resource availability and latitude. For many species, clutch size and/or fecundity is a function of body size, with larger organisms producing more offspring. Often body size is also a function of the age of the organism, with older individuals typically being larger. The text also presents an energetic context for considering reproductive systems that involve hermaphrotism and gender change. In some cases, gender change is a function of budgeting reproductive costs during the life cycle; the organism alternates between having a less expensive male reproductive biology at some times and a more costly female function at other times.

Ideas about reproduction become more complicated when one considers aspects of gender allocation and sex ratios (sex ratios were initially considered in Chapter 17). The question revolves around the cost of raising one gender versus the other. Models of gender allocation are related to parental condition and resource competition. Studies in support of each of these theories are presented.

The discussion of r-selection and K-selection incorporates many of the ideas previously presented in Chapter 20, such as timing of reproduction, number of offspring, resource availability, reproductive costs, etc. An understanding of the logistic growth model presented in Chapter 18 is also involved in r-selection and K-selection. One should realize that the concept of r- and K-selection is somewhat controversial, although often accepted as ecological dogma. To cite Dawkins: "Ecologists enjoy a curious love-hate relationship with the r/K concept, often pretending to disapprove of it while finding it indispensable" (See Boyce, 1984, in **Resource Materials**). At the very least, caution should be used in the presentation of this theory. As the textbook emphasizes, r and K strategies represent theoretical endpoints of a gradient of life history strategies. Some species are neither clearly r-selected nor K-selected, while other species appear to be one or the other under various environmental conditions. Although these strategies are applied to species many times, the safest alternative is to apply their usage to individuals. In plants, an alternative life history strategy is C-, S- and r-selection. In general, this system represents various plant responses to combined degrees of stress and disturbance.

Chapter 20 concludes with an important point about habitat selection by animals. Namely, if we are to manage and preserve species effectively, we must know the basis on which they choose particular habitats. The text provides many examples of criteria on which habitat selection may be made. The answers are not easy, however, since (1) apparently suitable habitats sometimes are not selected, (2) great plasticity occurs in habitat selection, and (3) habitat preferences vary over a species' range. Unfortunately, we have only a rudimentary understanding of this important facet of population biology.

In conclusion, much of Chapter 20 provides an ecological and evolutionary context for understanding reproductive biology. Although many theories and models have been advanced, good field data simply don't exist to ascertain the various facets of reproductive decision-making among species. It is important, nevertheless, that students gain an appreciation of the complexity of reproductive behavior and life history patterns.

TOPICS FOR DISCUSSION

1. Consider human parental investment in offspring. Approach investment from a variety of perspectives. For example, what are the demands of offspring on the time budget of parents? How do children affect the allocation of funds in a financial budget? Do children affect the energy budget of each parent? What kinds of activities require energy expenditures by parents? What is the effect of having one offspring versus multiple offspring on each of these budgets?

2. Using your answers to the previous questions as a starting point, evaluate the value of monogamy in humans to successful rearing of offspring? Does single parenting increase the fitness of the parent and/or the offspring?

3. Although controversial, do principles of mate selection in animals apply to the manner in which humans choose their mates? What are some similarities and differences? With regard to mate choice, compare the variation in human lifestyles to the natural variation exhibited in animal populations.

4. Why is mate choice among animals primarily a female choice? Can this be related to the respective costs of reproduction in females versus males? If so, how?

5. Is a consideration of the life history traits of individuals in a species relevant to the management/preservation of that species?

6. Why is the model of r- and K-selection too simplistic to be very realistic? Is the magnitude of K in a population necessarily resource-based? What are various life history traits that may determine or influence variation in K? Refer to Boyce (1984) in **Resource Materials** for further insights.

LECTURE OUTLINE

It's not an easy task to organize information about reproductive biology into a logical sequence to facilitate student understanding because so many of the ideas and theories overlap in a variety of contexts. The text presents a logical sequence and is essentially followed in the outline below. Thoughts about where to consider r/K selection are varied and could just as easily be included after consideration of the logistic growth model from Chapter 18. However, as Chapter 19 illustrates, maximizing reproductive effort and fitness is the core of understanding life history processes.

I. **Overview to Reproductive Biology**

    A. Asexual versus Sexual Reproduction
    B. Monogamy versus Polygamy
    C. Polygamy
        1. Polygyny
        2. Polyandry
        3. Mate Choice
            a. Resource-based Selection
            b. Genetically-based Selection

II. **Parental Investments and Energy Budgets**

    A. Iteroparity versus Semelparity
    B. Precocial versus Altricial Offspring
    C. Number of Offspring
        1. Influence of Age
        2. Influence of Body Size
        3. Resource Availability
    D. Sequential Gender Change and Reduced Energy Expenditure

III. **Models of Sex Ratio Adjustment (Gender Allocation)**

    A. Female Condition
    B. Local Resource Competition

LECTURE OUTLINE CONTINUED:

IV. Life History Models

    A. *r*- and *K*-selection
        1. Overview of *r* and *K* Strategists
        2. Constraints of Model
            a. Application to Species
            b. Fallacy of Deterministic Environments
    B. *C*-, *S*- and *r*-Selection in Plants

V. Basis of Habitat Selection

    A. Cues for Decision Making
        1. Resource Availability
        2. Habitat Structure
        3. Landscape Pattern
        4. Microclimatic Influences
    B. Plasticity in Habitat Selection
    C. Habitat Selection and Species Preservation

RESOURCE MATERIALS

Mating systems, mate choice and sexual selection have attracted much interest from ecologists and animal behaviorists; hence, an extensive literature exists. Alcott (1994) offers a particularly good textbook that addresses these topics (separate chapters are dedicated to presentation of mating systems, sexual reproduction and parental care, and male and female reproductive tactics). For specific details about reproductive ecology in mammals, consult Flowerdew (1987), which also contains a chapter about interactions between mammalian biology and life history parameters. Similarly, Gubernick and Klopfer (1981) offer a collection of papers dealing with many facets of parental care in mammals, including parental investment, parental aggression and sibling interactions. Rubenstein and Wrangham (1986) divide their book into two parts dealing with monogamous and polygynous mating patterns, which is particularly useful because of the many specific examples provided for birds and mammals. Information about mating and sex change in fish can be found in Shapiro (1994). Information about reproduction in plants can be found in a collection of papers provided by Wyatt (1992). Searcy (1982) reviews theory about how choice is made, who chooses whom and the reasons for mate choice (fertility, genetic quality, parental care, resource quality). Bateson (1983) offers separate chapters dedicated to non-random mating, mating decisions and life history tactics. Shaw and Darling (1985) provide an easy-to-read book that contains many good examples. Other information about the ecology of mate choice and sex can be found in books by Greenwood and Adams (1987) and Kirkpatrick (1987). Mock et al. (1990) summarize information and literature about avian siblicide.

The references presented in the textbook about specific life history strategies (*r*- and *K*-selection, and *C*-, *S*-, and *r*-selection) are particularly useful for lecture preparation. Refer to the citations to Grimes (1977), Boyce (1984) and Stearns (1976).

Alcock, J. 1994. *Animal Behavior* (5th Edition). Sinauer Associates, Inc., Sunderland, MA.

Bateson, P. (editor). 1983. *Mate Choice*. Cambridge University Press, Cambridge.

Flowerdew, J.R. 1987. *Mammals - Their Reproductive Biology and Population Ecology*. Edward Arnold, London.

Greenwood, P.J. and J. Adams. 1987. *The Ecology of Sex*. Edward Arnold Publishers, London.

Gubernick, D.J. and P.H. Klopfer (Editors). 1981. *Parental Care in Mammals*. Plenum Press, New York.

Mock, D.W., H. Drummond, and C.H. Stinson. 1990. "Avian Siblicide." *American Scientist* 78:438-449.

Kirkpatrick, M. 1987. "Sexual selection by female choice in polygynous animals." *Annual Review of Ecology and Systematics* 18:43-70.

Rubenstein, D. I. and R.W. Wrangham. 1986. *Ecological Aspects of Social Evolution: Birds and Mammals*. Princeton University Press, Princeton, NJ.

Searcy, W.A. 1982. "The evolutionary effects of mate selection." *Annual Review of Ecology and Systematics* 13:57-85.

Shapiro, D.Y. 1994. "Sex change in fishes - the how and why." In: *The Differences Between the Sexes*, R.V. short and E. Balaban (editors). Cambridge University Press, Cambridge. pp. 105-130.

Shaw, E. and J. Darling . 1985. *Female Strategies*. Walker and Company, New York.

Wyatt, R. (editor). 1992. *Ecology and Evolution of Plant Reproduction*. Chapman and Hall, New York.

CHAPTER 21

POPULATION GENETICS

CHAPTER SUMMARY AND ORGANIZATION

A rudimentary understanding of population genetics is essential to any beginning ecology student. On the one hand, a working knowledge of the genetic components of a population are essential from an applied viewpoint of habitat and species preservation. In a more theoretical context, however, such understanding is relevant to the processes of organismal evolutionary change and adaptation. As emphasized in Chapter 3, the subject of ecology is virtually meaningless without an evolutionary base. This chapter begins with a description of the basis for selective processes, i.e., genetic variation. Following sections cover the process of natural selection and various constraints on natural selection, respectively. Some instructors may want to combine this information with the presentation of adaptation in Chapter 3, or vice versa. Furthermore, since the topics of adaptation and natural selection should form the foundation for a contemporary ecology course, it might be desirable to incorporate them into early lectures or student discussions.

The primary focus of the first section is directed to the various mechanisms by which genetic variation arises in populations. Most of the concepts should be a review for students who have had a previous course in genetics or perhaps even introductory biology. Of the concepts presented in the first section, emphasis should be placed on the significant amount of genetic variation that results from gene recombination during sexual reproduction. This emphasis is needed to thwart the misconception that most students will have about mutations being the only important way of expressing new genotypes. New alleles can only rise from mutations, but observable variation among population members results primarily from the recombination of existing genetic information.

The second section of Chapter 21 presents an overview with examples of Hardy-Weinberg equilibrium. Students should understand the assumptions relevant to the existence of this equilibrium and that these assumptions are rarely met in natural populations. Mating is rarely random, populations are virtually never closed, population size is not infinite, etc; consequently, the genotypic frequencies predicted by the Hardy-Weinberg equilibrium equation are rarely found in nature. Nonetheless, this equilibrium can serve as a theoretical baseline for comparison in studies of populations. In addition, the Hardy-Weinberg equilibrium is essential to understanding coefficients of inbreeding presented later in the chapter.

Natural selection should be approached in the framework of selection for higher fitness among population members. Students should first understand that *fitness* is based on fecundity and leaving descendants that will themselves reproduce. The underlying determinant of fitness is an individual's genotype. Because great genetic variation exists among population members, selection has been visualized as a process that acts on the frequency of occurrence of genotypes over time. In stable habitats, the average expression of genotype/phenotype prevails and the frequency distribution of phenotypes appears as a normal distribution (stabilizing selection). In contrast, disruptive selection occurs in patchy habitats and results in selection for the extremes of phenotypic expression. The resulting frequency distribution is bimodal. The text provides a discussion of industrial melanism in the peppered moth as an example of disruptive selection. Directional selection is typically represented as a shift from dominance by an average phenotype to dominance by an extreme phenotype in a frequency distribution. One doesn't have to look far for examples of directional selection in contemporary life. Selection for increased pesticide resistance among pest insects and increased resistance to antibiotics among various pathogenic bacteria are two such examples. Stabilizing selection is undoubtedly the most common selective type on an ongoing basis and it occurs in all species; however, examples are fewer because no net phenotypic change is occurring. Species which have exhibited relatively little change over geologic periods are undoubtedly subject to stabilizing selection (e.g., horseshoe crabs). Such species are presumed to occur in stable habitats and are often of limited mobility.

Group selection increases the frequencies of genes that benefit a group of individuals (population), rather than benefiting single individuals which carry the gene. In fact, if the genes aren't selectively neutral, they can be detrimental to an individual that possesses the gene. This paradox is discussed in detail by the text and will need to be explained carefully to students. The main approach by the textbook is to consider group selection in terms of the evolution of altruistic behaviors, particularly among closely related, social species (kin selection). In this context, the idea of inclusive fitness is presented, whereby individuals can acquire additional fitness by elevating the fitness of close relatives. In a departure from the textbook, instructors may want to include possibilities for group selection other than altruism. Group selection often has been considered as one aspect of population regulation and growth in which genes may be selected if they prevent overpopulation (see Hartl and Clark, 1989 - **Resource Materials**).

One consequence of existing in small, localized populations is inbreeding. In most instances, inbreeding is detrimental to a population. If deleterious recessive alleles are present in the population, inbreeding will increase the expression of those alleles by maximizing levels of homozygousity. Inbreeding depression results from the negative effects associated with increased levels of homozygousity in a population. Emphasis should be placed on the idea that inbreeding results from nonrandom mating, i.e., the probability is higher that an individual will mate with a relative than a nonrelative. Inbreeding can be characterized by inbreeding coefficients, which are described in detail in the text. Students will need to be thoroughly familiar with the Hardy-Weinberg principle to understand these coefficients. Inbreeding coefficients are used by geneticists studying human populations, plant and animal breeders and conservation geneticists.

A second consequence of mating among individuals in a small population is genetic drift. Genetic drift results from random changes in gene frequencies due to random mating in the local population as compared to the global population. Genetic drift should be given thorough consideration in the classroom because it represents one potential mechanism leading to speciation, particularly drift in a population started by a small number of immigrants (founder's effect). One measure of genetic drift is the fixation index, which is based on differences between heterozygousity in a local population and heterozygousity in the global population. In determining the amount of genetic drift that might occur, the actual size of a local population may not be important; rather, the actual number of individuals contributing to reproduction is the crucial parameter. This has been termed the effective population size. Effective population size is influenced by the demographic characteristics of the population, most importantly the sex ratio and numbers of nonbreeding males. The effective population size is also affected by the neighborhood size of a population. The text defines neighborhood size as "the root mean square of dispersal distance of individuals about their natal origin." An alternative definition used by population geneticists is the area "around an individual within which the gametes that produced the individual can be considered to have been drawn at random" (Lincoln et al., 1982). No matter how neighborhood size is defined, the key idea is that the amount of dispersion from a birthplace is a determinant of effective population size.

As is the case with most chapters in the textbook, Chapter 21 concludes with a practical perspective. The theme of this section is the "minimum viable population" density, or the minimum size needed for a population to persist. In an environmental context, protection of endangered species requires knowledge about the minimum viable population sizes. The minimum viable population size for a species is closely coupled to its life history. Students should understand that it is not sufficient merely to sustain numbers of individuals; other important facets of the population also must be maintained. Consideration must be given to such obvious parameters as the age structure and sex ratio. Perhaps even more critically, however, losses of genetic variability because of death and dispersion must be compensated for by gains in genetic information. The text discusses the question of the minimum population size needed for the mutation rate to maintain pace with loss of genetic information.

TOPICS FOR DISCUSSION

1. The acquisition of adaptations, of course, is based on sufficient genetic variability in a population for environmental selection to occur. Are there consequences associated with extreme amounts of genetic variability? Does this mean that many individuals are not adapted or are suboptimally adapted to their surroundings? Instructors may want to investigate the idea of the "genetic load" associated with the genetic variations in populations.

2. Food for thought: A very controversial idea deals with "directed mutations." The basic idea is that cells have some control or are able to direct the kinds of mutations that occur in response to stressful situations (see Gillis, 1991, or McDonald, 1983, in **Resource Materials**). How does this situation compare with traditional ideas of evolutionary biology as presented in the textbook? If true, would this theory require the rethinking of concepts such as minimum viable population size?

3. What are the population/genetic consequences of human hunting practices in which the largest, most robust animals usually are taken as trophies? By the same token, how does the removal of aggressive individuals (such as bears) from parks or areas of human habitation affect genetic-based behavioral patterns in a population?

4. In a genetic context, what are the possible population consequences of the fragmentation of natural habitats by humans?

5. Refer to review questions 20 and 21 in textbook.

LECTURE OUTLINE

One theme for any ecological course must be the evolutionary basis for most ecological understanding. In this context, it might be advisable to integrate information from Chapter 21 with the information about adaptation in Chapter 3 early in a course. Clearly, species acquire adaptations because of environmental selection acting on existing genetic variation. The outline below encompasses this integrative approach; however, please also refer to the lecture outline for Chapter 3.

I. Basic Evolutionary Ideas and Definitions

    A. Fitness
    B. Adaptation
    C. Natural Selection
        1. Genetic Variation
            a. Role of Sexual Recombination
            b. Mutations
                i. macromutations
                ii. micromutations
        2. Inheritance
        3. Environmental Selection (Non-Random Reproduction)
    D. Evolution

LECTURE OUTLINE CONTINUED:

II. Modes/Forms of Natural Selection

    A. Directional Selection
    B. Disruptive Selection
    C. Stabilizing Selection
    D. Group/Kin Selection

III. Important Aspects of Population Genetics

    A. Hardy-Weinberg Equilibrium
        1. Assumptions
        2. Hardy-Weinberg Equation
    B. Inbreeding
        1. Consequences
        2. Measurement and Quantification - Inbreeding Coefficient
    C. Genetic Drift
        1. Definition and Significance
        2. Measurement and Quantification - Fixation Index
        3. Effective Population Size
        4. Neighborhood Size
    D. Minimum Population Size

RESOURCE MATERIALS

In recent years, the literature regarding population genetics has expanded rapidly as many modern tools from molecular biology are being used in ecological studies (e.g., refer to the journal *Molecular Ecology*). Mettler et al. (1988) and Hartl (1989) offer excellent introductory textbooks to population genetics. Real (1994) provides a collection of papers dealing with ecological genetics. Berry et al. (1992) integrate aspects of genetics and ecology, including a survey of contemporary genetic techniques used in population studies. A number of review articles also have been published in the *Annual Review of Ecology and Systematics* (see below). Particularly relevant to the topics of Chapter 21 is the article by Elstrand and Elam (1993), which summarizes information regarding the effects of small population size on inbreeding, genetic diversity, fitness and gene flow for plants. Similarly, Slatkin (1985) reviews models, methods and examples of gene flow studies, while McDonald (1983) discusses information about levels of heterozygousity and genetic variation in populations. Brandon (1990), in particular, offers much relevant background material for a general ecology course. Group selection is also examined in the works by Brandon (1990) and Wilson (1983).

Barton, N.H. and B. Charlesworth. 1984. "Genetic revolutions, founder effects and speciation." *Annual Review of Ecology and Systematics* 15:133-164.

Berry, R.J., T.J. Crawford and G.M. Hewitt. 1992. *Genes in Ecology*. Blackwell Scientific Publications, London.

Brandon, R.N. 1990. *Adaptation and Environment*. Princeton University Press, Princeton, NJ.

Charlesworth, D. and B. Charlesworth. 1987. "Inbreeding depression and its evolutionary consequences." *Annual Review of Ecology and Systematics* 18:237-268.

Ellstrand, N.C. and D.R. Elam. 1993. "Population Genetic Consequences of Small Population Size: Implications for Plant Conservation." *Annual Review of Ecology and Systematics* 24:217-242.

Gillis, A.M. 1991. "Can organisms direct their evolution?" *Bioscience* 41:202-205.

Hartl, D.L. and A.G. Clark. 1989. *Principles of Population Genetics* (2nd Edition). Sinauer Associates, Inc., Sunderland, MA.

McDonald, J.F. 1983. "The molecular basis of adaptation: A critical review of relevant ideas and observations." *Annual Review of Ecology and Systematics* 14:77-102.

Mettler, E.L., T.G. Gregg, and H.E. Schaffer. 1988. *Population Genetics and Evolution.* Prentice Hall, Englewood Cliffs, NJ.

Real, L.A. 1994. *Ecological Genetics.* Princeton University Press, Princeton, NJ.

Slatkin, M. 1985. "Gene flow in natural populations." *Annual Review of Ecology and Systematics* 16:393-430.

Wilson, D.S. 1983. "The group selection controversy: history and current status." *Annual Review of Ecology and Systematics* 14:159-187.

CHAPTER 22

INTERSPECIFIC COMPETITION

CHAPTER SUMMARY AND ORGANIZATION

To this point in the textbook, the population topics discussed previously have dealt with various aspects of the population ecology of single species. Chapter 22, however, is the first of six chapters in which interactions between species are considered. The introduction to Chapter 22 sets the stage for these chapters by providing an overview to the various species interactions: interspecific competition, mutualism, commensalism, amensalism, parasitism and predation. With the exception of commensalism and amensalism, each of the other topics has at least one chapter devoted to it. Chapter 22 also introduces students to the ideas of exploitative and interference competition, both of which are essentially based on resource access and efficiency of utilization.

The introduction to the topic of interspecific competition in Chapter 22 is a good transition to the topic of interspecies interactions. For example, students already will be familiar with facets of competition from their study of intraspecific competition (Chapter 19); furthermore, the classic mathematical model of interspecific competition (Lotka-Volterra model) is an extension of the logistic growth model (Chapter 18). Instructors should realize from the outset, however, that great debate exists among ecologists about interspecific competition. For example, the Lotka-Volterra model is based on the assumption that competing species are at equilibrium and thus their interactions are density-dependent. How often is this actually the situation, and are most species at some equilibrium density? Another question relates to the degree to which the intensity of interspecific competition varies as a selective force. Is the effect of competition pervasive and unrelenting, or does its intensity fluctuate with overall environmentally variability, which causes an ebb and flow of resource availability, particularly food availability? Finally, how strong is the evidence for interspecific competition from studies of natural populations? For information related to the latter question, instructors should pay particular attention to the surveys of Schoener (1983) and Connell (1983), as cited by the text.

Despite the arguments regarding interspecific competition, competition theory is a basic tenet of contemporary ecology. As an evolutionary vehicle, competition should select either for individuals that can dominate resources (superior competitors) or individuals that, by one mean or another, are able to avoid competition and its negative effects sufficiently. The topic of interspecific competition begins in Chapter 22 with a treatment of the classic Lotka-Volterra equations. Although this mathematical treatment initially may be intimidating to students, it is particularly valuable as a teaching tool. First, it helps students to appreciate and understand more complex applications of the logistic model. Second, the Lotka-Volterra model illustrates, both mathematically and graphically, the four outcomes of competition between two species: (1) species number one wins a competitive interaction by driving species number two to local extinction; (2) species two wins a competitive interaction by driving species one to local extinction; (3) either species could win depending on the environmental circumstances; or, (4) both species eventually coexist. Third, the models set the stage for later topics; in particular, the ability of one species to drive another to extinction is the focal point of the competitive exclusion principle. Just as important, however, is a thorough understanding of the fourth outcome: coexistence. Coexistence obviously defines the life of organisms; so, what competitive mechanisms shape coexistence?

The second section of the chapter provides an overview of various investigations of competition. This chapter covers much ground dealing with field versus laboratory studies and studies of coexistence versus studies of exclusion. Instructors may find some of these examples relevant to other topics discussed in the chapter and should feel free to explore alternative organizational formats. For instance, the description of Gause's experiments with *Paramecium caudatum* and *P. bursaria* serves as a good example of niche shift and resource partitioning. Furthermore, although allelopathy is presented only in this section, it serves as an excellent example of interference competition among plants. In addition to these examples, the description of diffuse competition should not be overlooked. Most of our ideas about competition are necessarily simplistic and approached from the standpoint of only two competing species, but many species compete simultaneously, which renders structuring of communities by competition far more complex.

Ideas about competitive exclusion, the niche, resource partitioning and coexistence (all of which are presented in the remaining sections of Chapter 22) are closely related. For example, competitive exclusion is related to the outcome of competition when two species have "identical ecological requirements." If species have identical requirements, then surely they must occupy essentially the same niche. Similarly, species only can coexist if they have evolved some means to partition available resources to the extent that competition is muted, i.e., their niches have changed through niche shift or niche compression. On this basis, one satisfactory organizational approach for a lecture is to avoid putting off consideration of the niche as the last topic. Rather, begin with an assessment of the niche and how niche expression can be related to competitive interactions. There is no need to belabor the numerous definitions of the niche; instead, simply focus on the environmental niche and how it can be related to competition. This initial discussion of the niche then can be used as a foundation for understanding mechanisms of resource partitioning and coexistence. For example, niche shift and compression are best presented in the context of the fundamental versus the realized niche, both subdivisions of Hutchinson's environmental niche. The point here is that, although separate sections are devoted to these topics in Chapter 22, instructors will need to integrate their lectures into a unified presentation from the standpoint of helping students to understand the relationships among these ideas.

In summary, the role of interspecific competition in affecting community structure is quite variable. Competition may or may not result in local extinction of a species, may or may not result in resource partitioning, may or may not be represented by niche overlap. Nevertheless, competitive interactions do occur and shape the existence of species. Consequently, a strong theoretical base exists for understanding and predicting competitive outcomes. Instructors should study Chapter 22 carefully to understand the interconnections among the various topics and to become aware of the many facets and assumptions related to competition theory. With this background, instructors then should consider the organizational approach best suited for lectures to their students.

TOPICS FOR DISCUSSION

1. Would you agree with the statement, "A species goal should be to avoid competition"? Why or why not? What are some of the costs of competition? Do winners of a competitive interaction experience costs; if so, in terms of competition theory, can an analogy be drawn between groups of humans (e.g., countries) that find themselves embattled in competitive struggles?

2. Under what conditions should intraspecific competition be more severe for members of a species than interspecific competition? What are some features of various life history strategies that are relevant to your answer? Does the dispersal capability of a species affect your answer?

3. Apply some of the ideas of classic competition theory, such as competitive exclusion, resource partitioning, and equilibrium population densities, to Grime's description of the life history strategies of plants (*C*-, *S*- and *r*-selection). For example, resource partitioning would be most important for which strategy? Would the degree of niche overlap be as important for an *r*-selected species as a *K*-selected species?

4. In a population management context, much emphasis has been placed on humankind's removal of predators and the response of prey populations. In this same context, consider the role of competitors. Does overexploitation of some species affect community structure by allowing the populations of competing species to expand?

5. In a competitive context, why do introduced, exotic species often outcompete native species? Good examples are the zebra mussel in the Great Lakes region and purple loosestrife in temperate, northern wetlands.

6. Refer to review question seven in the textbook.

LECTURE OUTLINE

As with most ecological topics, there are numerous ways to organize a presentation of interspecific competition. In the outline below, the sequence of topics is initially similar to the textbook. However, the niche concept is presented earlier for the purpose of using the ideas of niche shift and compression as a basis for discussing coexistence and resource partitioning.

I. Introduction to Species Interactions

    A. Competition
    B. Predation
        1. Carnivory
        2. Herbivory
        3. Parasitism
    C. Mutualism
    D. Commensalism
    E. Amensalism

II. Types of Competition

    A. Exploitative Competition
    B. Interference Competition
    C. Diffuse Competition

III. Lotka-Volterra Model

    A. Assumptions of Model
    B. Mathematical Interpretation
    C. Graphical Illustration of Competitive Outcomes
    D. Laboratory and Field Examples of Each Competitive Outcome
    E. Competitive Exclusion Principle

IV. Models of Plant Competition - Differential Resource Utilization

V. Competition and the Niche Concept

    A. Environmental Niche
        1. Fundamental Niche
        2. Realized Niche
    B. Niche Change
        1. Niche Compression
        2. Niche Shift

LECTURE OUTLINE CONTINUED:

       C. Niche Overlap
          1. Indicator of Competition
          2. Constraints on Use as a Competitive Indicator

VI. Resource Partitioning

       A. Examples
       B. Coexistence

RESOURCE MATERIALS

Instructors are especially encouraged to read the textbook's citations to David Tilman's work regarding the importance of resource ratios to competition theory. Good introductions to classic competition theory can be found in the works by Miller (1967) and Hassell (1976). In addition, refer to the citations for Hedrick (1987) and Real and Brown (1991) in the **Resource Materials** to Chapter 17. Grace and Tilman (1990) offer a valuable multi-authored work containing detailed information on many aspects of plant competition. The work by Harper (1977), cited in the textbook, also provides useful information relevant to plant competition. Information integrating the niche concept and competition can be found in the books by Giller (1984), Wallace (1987) and Pontin (1982). The two latter books also address coexistence and the consequences of competition in communities.

Giller, P.S. 1984. *Community Structure and the Niche*. Chapman and Hall, New York.

Grace, J.B. and D. Tilman (editors). 1990. *Perspectives on Plant Competition*. Academic Press, New York.

Hassell, M.P. 1976. *The Dynamics of Competition and Predation*. Edward Arnold, London.

Miller, R.S. 1967. "Pattern and Process in Competition." *Advances in Ecological Research* 4:1-74.

Pontin, A.J. 1982. *Competition and Coexistence of Species*. Pitman Publishing, Inc., Boston, MA.

Wallace, A. 1987. *The Niche in Competition and Evolution*. John Wiley and Sons, Ltd., New York.

CHAPTER 23

PREDATION

CHAPTER SUMMARY AND ORGANIZATION

Conceptually, predation is easy to comprehend: one organism meets its energetic needs by consuming another living organism. Numerous questions, however, arise with regard to the various strategies of predation, the reciprocal effects of predator and prey populations on one another, the impact of predation on the diversity of community assemblages, etc. These topics are explored in chapters 23-26. Chapter 23 sets the stage for considering specific forms of predation (herbivory, carnivory and parasitism) in the later chapters by providing an overview to general predation theory. The strategy of this introductory chapter is to demonstrate that predation is more than a mere transfer of energy between organisms/trophic levels; rather, predation is a complex interaction between predatory and prey species, each clearly influencing the evolution of the other.

The discussion of predation begins with a presentation in chronological order of four models of predation. These models predict the changes in predator and prey population size that would occur because of the predator-prey interaction. The first model, the Lotka-Volterra model, predicts regular oscillations of predator and prey population densities; however, the model is based on overly simplistic assumptions. A later model, the Nichols-Bailey model, was developed for parasitism. This model predicts the same end result of oscillation between predator and prey densities, but the underlying assumptions are quite different (prey uniformly distributed versus randomly distributed, discrete time units versus continuous, delayed conversion of energy into offspring versus immediate conversion). The Rosenweig-MacArthur model is a more recent, somewhat more realistic model. The model differs from the previous models in that non-linear growth responses of prey and predators are possible, as shown in Figure 23.3 of the textbook. On the basis of this model, predator-prey interactions can result in either stable population cycles, unstable cycles or damped changes in density. If only one model is to be emphasized in the classroom, the Rosenweig-MacArthur model should provide students with the largest conceptual framework in which to consider predation. A final model is briefly presented in which ratios of prey to predator density are important. In this model, predator and prey density are a function of prey productivity.

The central component in each of the models presented above is the potentially cyclic behavior of predator and prey density patterns. The second section of the chapter regarding predator-prey systems explores this idea. In general, cyclic predator-prey density patterns arise only when specific conditions are met. These conditions include heterogeneity of the habitat, prey immigration and prey dispersion patterns. Examples of classic studies related to these ideas are presented in detail. In essence, prey populations must be able to recover from predation or face extinction. Similarly, predators cannot withstand prolonged periods of low prey density.

Prey density influences most aspects of a predator's existence, and this relationship should be emphasized in the classroom. Typically, predators respond to increasing prey density in two ways: the functional and numerical responses. The functional response expresses the relationship between predation rate and prey density and is categorized into three models. Each model demonstrates greater consumption rates at higher prey densities, but with significant differences.

Students should be thoroughly familiar with these models and their underlying assumptions. The type I model assumes that consumption rate increases linearly with increasing prey density until predator satiation is reached. A type II functional response involves a deceleration of consumption rate to a plateau due to the phenomenon of "handling time." A type III functional response is the most complex and can involve learning behavior, switching between prey items and the establishment of a search image by a predator. The textbook also uses the type III model to address the threshold of security in prey.

The numerical response is based on the idea that the density of predators will change in response to prey density. These changes may occur because of a change in predator emigration/immigration patterns or may result from an increase/decrease in rates of predator mortality and natality. Although the textbook presents the functional and numerical responses after a discussion of the predator-prey mathematical models, an alternative organizational approach would be to present the predator responses first. The reason for this approach rests with the fact that changes in predator density as a function of prey density are implicit to mathematical predator-prey models; furthermore, increases in predator density typically result from higher consumption rates (i.e., positive change in functional response) and the associated changes in mortality and natality. Hence, it may make intuitive sense for students to grasp the easily understood ideas of functional and numerical response first before exploring the mathematical models.

Chapter 23 concludes with a detailed presentation of optimal foraging theory. In a general way, the basic tenets of optimal foraging (optimal diet and marginal value theorem) relate to rules of decision-making that one would expect predators to follow with regard to prey choice and foraging efforts. These rules are presented in the textbook with example studies and should fall into the realm of common sense for students. Students also should understand, however, that predators are constrained by and respond to other aspects of their environment to the extent that strict adherence to these decision-making rules is unlikely. While these rules may generally apply, most consumers are opportunists and may not always forage in the most optimally efficient manner. For example, the amount of acceptable risk to take while foraging is one key component of decision making for a predator. The text describes two forms of risk for a predator: "risk- sensitive foraging" and "predation risk." Risk-sensitive foraging relates to the risk associated with visiting a foraging site of uncertain quality, i.e., the predator runs the risk of low food availability. Generally speaking, it does not make sense to run this risk when food resources support an adequate energy budget. In contrast to this form of risk, predation risk is associated with the possibility of the predator becoming prey for some other consumer; hence, protection has to be one component of making decisions about where to forage.

Each of the ideas presented in this chapter are conceptually important to understanding predator behavior and should not be overlooked in a general ecology course. These concepts should be viewed as guidelines for understanding the nature of predator-prey interactions with an understanding that predator and prey coexist in a complex environment, thereby making absolute predictions impossible. In an evolutionary context, acquisition of energy and nutrients is a basic need for all organisms and is closely coupled to reproduction and survival. Furthermore, the evolutionary significance of predator-prey interactions cannot be understated. All species are part of an evolutionary system in which prey continually evolve tactics and defenses to reduce predation and predators respond with their evolutionary counter-strategies. These are the topics of chapters 24-26.

TOPICS FOR DISCUSSION

1. What are some ways that humans, as predators, function as a selective force on prey? Do humans essentially follow the rules of optimal foraging? Why or why not?

2. Using the concept of optimal foraging and energy costs during searching, would a predator be more likely to control a prey population of high density or of low density? Would the predator be

more likely to control homogeneously dispersed prey or a prey that occurs in far-ranging, isolated clumps?

LECTURE OUTLINE

I. Predator Response to Prey Density

    A. Numerical Response
        1. Emigration
        2. Increased Natality
        3. Decreased Mortality
    B. Functional Response
        1. Type I Response
        2. Type II Response
            a. Handling Time
            b. Disk Equation
        3. Type III Response
            a. Switching
            b. Search Image
            c. Prey Threshold of Security
    C. Predator Aggregation

II. Models of Predator-Prey Interactions

    A. Lotka-Volterra Model
        1. Mathematical Formulation
        2. Assumptions of the Model
        3. Graphical Analysis of the Model
            a. Predator and Prey Isoclines
            b. Stable Oscillation
    B. Nicholson-Bailey Model
        1. Mathematical Formulation
        2. Assumptions of the Model
    C. Rosenweig-MacArthur
        1. Comparison to Lotka-Volterra Model/Isoclines
        2. Outcomes of Model
            a. Stable Population Cycles
            b. Unstable Population Cycles
            c. Absence of Population Cycles

III. Optimal Foraging Strategy

    A. Optimal Diet
        1. Rules for Decision Making
        2. Example Studies
    B. Marginal Value Theorem
        1. Rules for Decision Making
        2. Example Studies
    C. Risk-Sensitive Foraging
        1. Expected Energy Budget Rule
        2. Predation Risk

RESOURCE MATERIALS

Refer to the **Resource Materials** for Chapters 24 and 25 for further sources of information about predation. Several general references are listed below. Pianka (1994) summarizes much of the same information as the textbook regarding predator-prey models and predator responses. Pyke et al. (1977) and Pyke (1984) provide excellent reviews of optimal foraging strategy. Pfeffer's book (1989) is essentially a dictionary of predators with summaries of their life history and hunting tactics. Consult chapters three and four of the text by Crawley (1992) for further discussion of predator-prey population dynamics and foraging theory, respectively. Since the textbook doesn't address predation in aquatic communities, refer to the books by Zaret (1980) and Kerfoot and Sih (1987) for information.

Ratio-dependent predator-prey theory has received much attention in the recent literature. For example, the work by Matson and Berryman (1992) represents a special feature about ratio-dependent predation theory published in *Ecology;* related papers have been written by Gleeson (1994) and Ginzburg and Akcakaya (1992).

Crawley, M.J. 1992. *Natural Enemies: The Population Biology of Predators, Parasites, and Diseases*. Blackwell Scientific Publications, Oxford.

Ginzburg, L.R. and H.R. Akcakaya. 1992. "Consequences of ratio-dependent predation for steady-state properties of ecosystems." *Ecology* 73:1536-1543.

Gleeson, S.K. 1994. "Density dependence is better than ratio dependence." *Ecology* 75:1834-1835.

Holdt, R.D. and J.J. Lawton. 1994. "The ecological consequences of shared natural enemies." *Annual Review of Ecology and Systematics* 25:495-520.

Kerfoot, W.C. and A. Sih (editors). 1987. *Predation: Direct and Indirect Impacts on Aquatic Communities*. University Press of New England, Hannover, NH.

Matson, P.A. and A.A. Berryman (editors). 1992. "Special Feature: ratio-dependent predator-prey theory." *Ecology* 73:1539-1566.

Pfeffer, P. 1989. *Predators and Predation*. Facts on File, Oxford.

Pianka, E.R. 1994. *Evolutionary Ecology* (5th Edition). HarperCollins College Publishers, New York.

Pyke, G.H., H.R. Pulliam, and E.L. Charnov. 1977. "Optimal foraging: a selective review of theory and tests." *The Quarterly Review of Biology* 52:137-154.

Pyke, G.H. 1984. "Optimal foraging theory: a critical review." *Annual Review of Ecology and Systematics* 15:523-575.

Zaret, T.M. 1980. *Predation and Freshwater Communities*. Yale University Press, New Haven, NH.

CHAPTER 24

VEGETATION-HERBIVORE SYSTEMS

CHAPTER SUMMARY AND ORGANIZATION

A first impression of herbivory can be misleading. Plants seem to be the perfect prey; they can't hide, their lack of mobility prevents active evasion of predators, and the overall abundance of plants would suggest a lush, green world ready for consumption. In reality, the inability to escape herbivores directly has resulted in an intense, often subtle, evolutionary system involving plants and herbivores. Thus, Chapter 24 deals with the significance of herbivory from both the perspective of the plant and the herbivore. When considering herbivory, students should understand that herbivory differs significantly from carnivory, in which the prey is almost always killed. Grazers often function essentially as parasites, consuming and damaging portions of the plant without causing plant death. In other instances, herbivory of seedlings for example, death is probable. Further contrasts to carnivory relate to herbivore specialization on specific plant parts, sometimes eating only leaves, flowers, fruits, seeds or, less often, woody structures. Depending on the plant part consumed, its age and condition, the impact of herbivory may or may not be significant.

The first major section of Chapter 24 covers the effect of herbivory on plant fitness. Even without causing death, herbivory can decrease plant fitness. The text provides insight into many reasons fitness declines with herbivory. Some of these reasons include increased mortality, lower reproductive effort, loss of biomass or nutrients and reduced competitive ability. Overall plant vigor often declines, particularly if plant reserves are drained during the replacement of consumed tissues. Coupled with these negative impacts are physiological anomalies, (e.g., hormonal regulation), which can be induced by significant defoliation. Negative impacts on fitness are generally greatest when propagules, reproductive structures and young tissues are consumed. In this section, the textbook also provides information about stimulation of plant growth by modest grazing pressure; although flushes of plant growth/reproduction can be associated with mild grazing, no conclusive evidence exists to suggest that long-term plant fitness is elevated.

In the second section of the chapter, fitness is explored from the point of view of the herbivore. In fact, plants may affect the fitness of herbivores more than herbivores affect the fitness of plants. A key point of emphasis relates to the quality of plant material versus quantity of plant material available for consumption and how this differs from carnivore systems. Relative to the availability of plant material, the prey items for carnivores are usually less abundant, but of reasonable quality. In contrast, food availability to herbivores is high, but the quality is often low. Hence, fitness in herbivores is more influenced by food quality than by the costs of finding, killing and handling prey, as in the case of carnivore systems. The overall quality of plant tissues to herbivores is influenced by a number of factors: tissue age, cellulose content, lignin content and the presence of digestive-reducing compounds, (e.g, tannins), as well as the presence of other secondary plant chemicals that may make the plant less palatable. Plant secondary compounds also may reduce the fitness of herbivores by causing hormonal imbalances and affecting reproductive processes.

Although plants may seem outwardly passive, they have evolved a variety of strategies to discourage herbivory. Among members of the plant kingdom, at least four main defense strategies can be found. First, some plants mimic other plants or even some aspect of their predators, either of which serves to suppress herbivory. Second, structural defenses have been hypothesized, but conclusive experimental evidence is weak. Possible physical barriers include hard structures, such as thick seed coats, spines, and pubescence. A third strategy occurs when seed production is synchronized over a very short period of time. In such cases, seed production exceeds predator demand; hence, predators are able to eat to satiation (termed "predator satiation") and some portion of the seed crop escapes predation. The fourth general defensive strategy involves the production of secondary chemical compounds. These chemicals, their effects and manner of release are quite varied among plant examples, but they all have the common purpose of deterring herbivores. These compounds may be transported to wound sites to suppress infections or they may accumulate as general inhibitors to prevent herbivore attacks. Numerous examples of plant chemical defense are described in the text.

Whether or not these secondary chemical compounds in plants actually evolved for defensive purposes or for other reasons is an open question. Nonetheless, they do serve as a powerful selective force on the evolution of counter-strategies by herbivores. Counter-strategies run the gamut from selection for specific feeding behaviors to possession of internal detoxifying systems. Mixed function oxidase (MFO) is one detoxification mechanism possessed by all animals that certainly helps herbivores adjust to plant toxins. In general, highly-evolved counter-strategies are typical of herbivore feeding specialists. In some instances, the herbivore is so closely co-evolved with its plant prey that the herbivore enhances its own defensive capabilities by accumulating the plant's toxins.

Chapter 24 concludes in a theoretical context. The question of functional response in herbivores is first addressed followed by a discussion of plant-herbivore models. Clearly, functional response between herbivores and carnivores is dramatically different. Herbivory requires a significant component of handling time, which involves grazing, chewing and swallowing. In addition, the spacing and apparency of plants affects herbivore search strategies. If the rate of forage consumption responds to and is limited by forage availability, the functional response is a classic type II curve. A positive numerical response by herbivores (increased herbivore density) can result in oscillations of herbivore density and vegetative biomass.

Chapter 24 provides an excellent overview to herbivory, however, discussion of herbivory in aquatic systems is not included. In the open water, herbivores include species of fish and zooplankton. Not to be overlooked, many invertebrates graze on the periphyton of littoral and benthic communities. A "scraping" mode of existence is common to a great many of these species whereby algae, and associated fungi and bacteria, are scraped from rocks, plants and other solid surfaces. Scrapers are an important component of the energetics of lotic communities. In lentic communities, considerations of herbivory recently have been considered in an applied context of attempting to regulate or manage these systems (refer to **Resource Materials** provided for Chapter 15).

TOPICS FOR DISCUSSION

1. What are some ways that herbivores may affect the competitive abilities of plants?

2. In an evolutionary context, compare the behavioral, morphological and physiological characteristics needed by herbivores as opposed to carnivores.

3. Humans often sow monoculture populations of crop plants in which the individuals are very evenly dispersed. Does this particular means of cultivation make sense in terms of available information about herbivory? Why or why not? Does this strategy increase or decrease the need to use pesticides? Why?

4. What kind of functional response curve might filter-feeding zooplankton exhibit? What are the arguments against a type III functional response curve for these organisms?

LECTURE OUTLINE

I. Introduction to Herbivory

    A. Forage Quality versus Forage Quantity
    B. Partial Consumption: Parasitism?

II. Influence on Plant Fitness

    A. Negative Effects
        1. Plant Death
        2. Removal of Tissue Only
        3. Plant Compensation
    B. Benefits: Increased Fitness?

III. Herbivore Fitness

    A. Food Quality
    B. Effect of Plant Secondary Compounds on Herbivore Reproduction

IV. Defense Strategies and Herbivore Counter-Strategies

    A. Deterrents to Herbivory
        1. Plant Density and Dispersion
        2. Plant Mimicry
        3. Plant Structures
        4. Plant Secondary Compounds
        5. Synchronized Seed Production
    B. Herbivore Counter-Strategies
        1. Feeding Behavior
        2. Detoxification Capabilities

V. Plant-Herbivore Relations
    A. Functional Response
    B. Numerical Response
    C. Plant-Herbivore Cycles

RESOURCE MATERIALS

A key reference that summarizes the effects of herbivores on plant populations and communities is offered by Huntly (1991). Most monographs about plant-insect or plant-animal interactions include discussions of herbivory. For example, chapters five and six in Harborne's book (1988) are dedicated to feeding preferences of insects and vertebrates, respectively, while chapter seven explores coevolution between plant defense and animal response. Similarly, chapters four and five in the book by Abrahamson (1989) deal with herbivorous insects and mammalian herbivores, respectively. Hunter et al. (1992) examine the importance of natural variability to plant-animal interactions (four chapters specifically dealing with herbivory). Volume 42, issue #1 of *Bioscience* deals specifically with the phylogeny of plant-animal interactions; most of these papers consider mutualism, but one paper is about host-use patterns by moths. Other works dealing specifically with phytophagous insects are those by Bernays (1994) and Strong et al. (1984). Karban and Myers (1989) summarize the changes that occur in plants because of herbivory.

The function of secondary plant compounds as antiherbivore defenses also has been explored in a number of monographs and papers. Bryant et al. (1991) provide an important review paper. Rosenthal and Berenbaum (1991) and Harborne (1988) offer particularly useful references. The effect of plant chemistry also includes elemental concentration, particularly nitrogen (refer to Mattson, 1980, cited by text; Scriber and Slansky, 1981).

Abrahamson, W.G. (editor). 1989. *Plant-Animal Interactions*. McGraw-Hill Book Company, New York.

Bryant, J.P., F.D. Provenza, J. Pastor, P.B. Reichardt, T.P. Clausen, and J.T. du Toit. 1991. "Interactions between woody plants and browsing mammals mediated by secondary metabolites." *Annual Review of Ecology and Systematics* 22:431-446.

Harborne, J.B. 1988. *Introduction to Ecological Biochemistry*. Academic Press, New York.

Hunter, M.D., T. Ohgushi, and P.W. Price. 1992. *Effects of Resource Distribution on Animal-Plant Interactions*. Academic Press, Inc., New York.

Huntly, N. 1991. "Herbivores and the dynamics of communities and ecosystems." *Annual Review of Ecology and Systematics* 22:477-503.

Karban, R. and J.H. Myers. 1989. "Induced plant responses to herbivory." *Annual Review of Ecology and Systematics* 20:331-348.

Scriber, J.M. and F. Slansky. 1981. "The nutritional ecology of immature insects." *Annual Review of Entomology* 26:183-211.

Strong, D.R., J.H. Lawton, and R. Southwood. 1984. *Insects on Plants: Community Patterns and Mechanisms*. Harvard University Press, Cambridge, MA.

CHAPTER 25

HERBIVORE-CARNIVORE SYSTEMS

CHAPTER SUMMARY AND ORGANIZATION

To some scientists, carnivores are considered true predators in the sense that the prey are animals that are killed quickly during an attack. Whether or not one finds this narrow definition of predation satisfactory, no doubt exists that the existence of a carnivore is fundamentally different from a herbivore's existence. In particular, predators must be well-adapted to finding prey of lower density (relative to plants) and must have the appropriate attack properties to overcome effective prey defensive strategies. As emphasized in the first section of Chapter 25, the fitness of a predator is determined by "its ability to capture prey." In this scenario, the carnivore and its prey are closely coupled in an evolutionary game in which prey fitness depends on escaping predation and predator fitness depends on finding prey. The first two sections of Chapter 25 are dedicated to exploring the outcomes of this evolutionary struggle. The first portion of the chapter provides an analysis of various defensive strategies exhibited by prey. The second section deals with the various offensive strategies utilized by predators.

Seven strategies to avoid being eaten are discussed in the text: chemical defenses, warning coloration, mimicry, cryptic coloration, behavioral defenses and predator satiation (warning coloration and mimicry are discussed together). Each of these should be given due consideration in the classroom. Always keep in mind that some defensive mechanisms make equally effective offensive tactics, such as the use of poisonous chemicals or cryptic coloration. The material in the text sets the stage for introducing other ideas in lecture. For example, the use of warnings by particularly toxic or pugnacious prey does not have to be limited to coloration; rather, warning sounds can be issued (consider rattlesnakes). Lecturers may want to be cautious in their discussions of Batesian and Mullerian mimicry. Mimicry involving the distasteful monarch butterfly and the presumably palatable viceroy butterfly has historically been used as the classic example of Batesian mimicry and is so described in the text. However, Ritland and Brower (1991) found that both the viceroy and monarch butterflies were distasteful to avian predators, thereby suggesting that the relationship between these taxa is actually Mullerian mimicry. Although only brief discussions are presented regarding cryptic coloration, behavioral defenses and predator satiation, these sections are intuitive and one doesn't have to look far to find numerous examples. Of these three topics, "predator satiation," which involves synchronous production of many offspring by prey, may require the most elaboration for students. Synchronous production of offspring can confuse predators, limit the proportion of the population consumed and may aid in defense of prey.

Predators have evolved various hunting strategies to insure profitable foraging in an energetic context. Some predators "sit and wait," while others stalk or pursue their prey. Each of these strategies is often accompanied by various adaptations. For example, a predator that primarily ambushes prey may be well-camouflaged with cryptic coloration. In some instances, predators may exhibit a unique form of mimicry entitled "aggressive mimicry," whereby the predator mimics its prey to the extent that it can go undetected long enough to exploit the prey. Other specific adaptations for enhancing the probability of prey capture include speed, claws, and well-developed teeth and jaw structure structures, as well as heightened sensory perception, such as exceptional hearing and eyesight.

Cannibalism (intraspecific predation) is considered in the third section of Chapter 25. This is a topic seldom given much thought by the average student or much treatment in contemporary ecological textbooks. Although cannibalism is evolutionarily disadvantageous, students should realize that cannibalism occurs among a diverse array of animal taxa. Under normal conditions cannibalism is probably not common in most of these taxa, but increases in frequency when populations are suffering from food shortages and stress. In some instances, cannibalism can be a significant form of predation (note the example of cannibalism in wall-eye populations described in the text). Short-term advantages associated with cannibalism result from reduced intraspecific competition and greater food availability for survivors. Another advantage may include recoupment of energy invested in offspring when parents eat their own progeny.

Intraguild predation is a unique interaction that includes facets of both predation and competition. Intraguild predation occurs when two predators are at the same trophic level (consume the same prey items), but one of the predator species also preys on the other predator. Clearly, the surviving predator removes a competitor while gaining an energetic and nutritional return from consumption of that competitor. In addition to providing examples, the textbook discusses the importance of intraguild predation to resource management, with an emphasis on the need to understand trophic structure, particularly when new species are to be introduced into an ecosystem.

Questions of predator regulation of prey density and the coupling of regulatory effects to predator-prey cycles are very difficult to answer. Attempts to investigate answers to these questions have been rendered more difficult because of the disruption of normal predator-prey dynamics by human activity. In other words, humans have so altered habitats, species numbers, and population densities that it is nearly impossible to find a location for the study of natural predator-prey interactions. Most students will be biased toward an idea that predators regulate prey populations and thereby force cycles of predator and prey density. The field studies described in Chapter 25, however, indicate that this is not always the case. For example, hare cycles in the boreal forest are driven by climatic cycles which are affected by sunspot activity. South of the boreal region, cyclic change in hare density is primarily dependent on the nature of the environment; hare populations rise and fall in uniform habitats, but are stable in patchy regions because of the combined effects of predation and hare dispersal between low- and high-quality areas. In contrast to these examples, predators can be the driving force for cycles of density for shrews and microtine voles. Much of the information in this section can be used as an effective supplement/contrast to the models of predation presented in Chapter 24. For predators to regulate prey populations, reproductive individuals must be consumed, which causes a decrease in r (rate of increase) of the prey population. The text provides examples in which predation provides some degree of prey regulation. In some instances, regulation occurs only when prey:predator ratios are low. Lecturers may want to consider developing a list of factors that influence the extent to which a predator may or may not control prey species. Consider, for example, prey defenses, prey distribution, prey life history, environmental heterogeneity, predator-prey ratios, etc.

Humans act as specialized predators who exploit populations while simultaneously attempting to manage those populations. The concluding section to Chapter 25 considers models and ramifications of human exploitation in detail. First, the concepts of standing crop, sustained yield and maximum harvest are presented. These topics are followed by a presentation of three management models currently in use: the fixed quota model, the harvest effort model and the dynamic pool model. These models are based primarily on the logistic growth model and have many limitations. In terms of biological considerations, such models often fail to incorporate critical information about the managed population, e.g., life history phenomena in conjunction with size and age classes, sex ratios, survivorship, environmental uncertainty, etc. More importantly, however, the models fail to include the consequences of exploitation as a commercial endeavor. Ultimately, the development of an industrial and economic base must be sustained by ever increasing exploitation. Clearly, such a system eventually is doomed to failure. Other problems also arise from failing to consider the role of exploited species in their ecosystem; unanticipated effects occur throughout the ecosystem from the overexploitation of one species. In the final analysis, the exploitation of species often is driven by the human desire for financial gain, usually without any real consideration of ecological consequences.

TOPICS FOR DISCUSSION

1. What might be some potential advantages/disadvantages of a "sit and wait" predatory strategy versus "active pursuit" for endotherms versus ectotherms? Consider this questions in terms of the metabolic requirements of ectotherms and endotherms.

2. Under what conditions might biological control be a realistic tool for managing a nuisance species? Consider attributes of the prey and predator as well as characteristics of the habitat. What are some examples in which biological control has been used effectively?

3. Would a predator be expected to specialize upon and regulate a prey species when the prey density is very low for reasons not coupled to predation? Why or why not?

4. Waterfowl and deer are two commonly hunted species; however, these groups have experienced dramatic changes in their density in recent times: deer have increased in density while waterfowl populations have declined. Essentially, the population density of these species has changed for the same overall reason of altered landscape patterns due to human activity. Should the philosophy of human exploitation differ for these two groups? How should management/hunting strategies differ between these two animal groups? Should hunting and management of these species be dependent on the life history characteristics and habitat requirements of these species?

5. Refer to review questions 8-11 in the textbook.

LECTURE OUTLINE

I. Carnivory

    A. Comparisons to Herbivory
    B. Predator and Prey: Evolutionary Consequerces
    C. Foraging Strategies: Advantages-Disadvantages
        1. Sit and Wait
        2. Trap
        3. Pursuit

II. Tactics to Avoid Being Eaten

    A. Cryptic Coloration
    B. Mimicry
    C. Aposematic Coloration
    D. Chemical Defenses
    E. Armor and Weapons
    F. Behavioral Adaptations
        1. Aggressive Behavior
        2. Alarm Calls
        3. Distraction Displays
        4. Prey Groups
    G. Synchronous Offspring Production

LECTURE OUTLINE CONTINUED:

III. Adaptations for Successful Predation

    A. Aggressive Mimicry
    B. Camouflage
    C. Offensive Weapons
        1. Teeth
        2. Muscular Jaws
        3. Claws
        4. Speed
    D. Prey Detection
        1. Hearing
        2. Vision
        3. Electrical Fields (Aquatic Habitats)

IV. Specialized Forms of Predation

    A. Interspecific Predation (Cannibalism)
    B. Intraguild Predation

V. Questions of Predator Influence on Prey Populations

    A. Do Predators Regulate Prey Populations?
    B. What are the Causes of Predator-Prey Population Cycles?

VI. Utilization and Management of Animal Populations by Humans

    A. Sustained Yield
        1. Terminology: Maximum versus Optimal Sustained Yield
        2. Mathematical Formulations
            a. General Equation
            b. r-selected Species
            c. K-selected Species
            d. Assumptions and Problems
    B. Management Models
        1. Fixed Quota
        2. Harvest Effort
        3. Dynamic Pool Model
        4. Shortcomings of Models

RESOURCE MATERIALS

Many of the other references cited in the **Resource Materials** in previous chapters are useful here. In particular, refer to Chapter 17 (Real and Brown, 1991; Hedrick, 1984), Chapter 18 (Tamarin, 1987) and Chapter 19 (Hassell, 1976; Zaret, 1980; Kirchman and Sih, 1987; Alcock, 1994) for information and chapters about predation/carnivores. Alcock (1994), in particular, includes chapters on the ecology of feeding behavior and antipredator defenses. With regard to defenses, a diverse literature exists about mimicry (e.g., Pasteur, 1982; Nihjout, 1994); Ritland and Brower (1991) clarify the mimicry between the viceroy and monarch butterflies. Taylor (1984) offers a good general introduction to predator-prey theory with chapters on predator functional response, predator-prey cycles and systems, etc. The work by Polis and Myers (1989) is cited in the textbook and is good for its consideration of the effect of intraguild predation at the individual, population and community level. Fox (1975) reviews field experiments of cannibalism and its effects on demographic structure and population processes. A good example of handling time involving sparrows and seed size is presented by Pulliam (1985).

Alcover, J.A. and M. McMinn. 1994. "Predators of vertebrates on islands." *Bioscience* 44:12-18.

Fox, L.F. 1975. "Cannibalism in natural populations." *Annual Review of Ecology and Systematics* 6:87-106.

Nihjout, H.F. 1994. "Developmental perspectives on evolution of butterfly mimicry." *Bioscience* 44:148-157.

Pasteur, G. 1982. "A classificatory review of mimicry systems." *Annual Review of Ecology and Systematics* 13:169-199.

Polis, G.A. and C.A. Myers. "The ecology and evolution of intraguild predation: potential competitors that eat each other." *Annual Review of Ecology and Systematics* 20:297-330.

Pulliam, H.R. 1985. "Foraging efficiency, resource partitioning, and the coexistence of sparrow species." *Ecology* 66:1829-1836.

Ritland, D.B. and L.P. Brower. 1991. "The viceroy butterfly is not a Batesian mimic." *Nature* 350:497-498.

Taylor, R.J. 1984. *Predation*. Chapman and Hall, New York.

Wickler. W. 1968. *Mimicry in Plants and Animals*. World University Library, London.

CHAPTER 26

PARASITISM

CHAPTER SUMMARY AND ORGANIZATION

For such a pervasive biotic interaction, parasitism usually is not given much attention by ecologists and ecological textbooks. Parasitism is not an easy interaction to define, but in the broadest sense, nearly 100% of all plants and animals are parasitized (see Esch and Fernandez, 1993 - **Resource Materials**). From an ecological point of view, several key aspects of parasitism can be emphasized initially. First, parasitism is detrimental to the host (some parasitologists might not agree with this assertion). Second, parasites are heavily dependent on their hosts, often with an intimate symbiotic relationship existing between the host and parasite. Third, many examples of parasitism serve equally well as examples of coevolution, a term which is often misused. The latter two points, perhaps, relate to the most intriguing facet of many parasitisms: the habitat in which a parasite often spends most of its life is alive! Consequently, hosts may function simultaneously as habitats and sources of energy and nourishment (this generalization would not apply to social parasitism). Various examples of parasitic effects on humans are also noteworthy (Lyme disease, Bubonic plague, venereal diseases, AIDS, etc.).

The first two sections of Chapter 26 provide an overview to various parasitic taxa, ways of describing parasites, and the role of hosts as habitat. Parasites may live on the inside or outside of hosts, (endoparasites and ectoparasites) and may be microscopic or macroscopic (microparasites and macroparasites). The third section deals with parasite life cycles; however, the emphasis is not so much on various examples of life cycles as on commonalities that can apply across a wide range of parasitisms. Modes of parasite transmission (direct and indirect) are described. The information in this section forms a sound basis for emphasizing coevolutionary aspects of parasitism. The life cycles of parasites have evolved to maximize transmission and invasion of hosts. The dynamics of transmission are particularly important in an ecological context and relate to the dispersion pattern of both the host and parasite. Hosts often are considered as analogous to islands from which parasites must escape and as islands that parasites must eventually invade. Parasite transmission is most effective in dense host populations. Parasites often illustrate an overdispersed frequency such that only a few host individuals act as parasite reservoirs.

A brief section also is presented regarding defensive responses and symptoms of infection in parasitized hosts. The primary means of defense involve either immune/biochemical responses or growth responses designed to sequester the parasite (e.g., cyst formation). The symptoms of parasitism are numerous and often specific to the parasite infestation. In some cases, symptoms may have a dramatic effect on host fitness. In particular, infestations may cause sterility, affect mate selection or result in abnormal, detrimental behavior.

The fifth section of Chapter 26 deals with various aspects of the population biology of parasites and their hosts. This section is organized around four themes of host-parasite population models: host density regulation, the dynamics of parasite growth, distribution response and evolutionary response. Clearly, each of these ideas can be related to other topics covered in the text. For example, (1) the model of host population change is merely a modification of the differential form of the exponential growth equation with inclusion of an expression for parasite-induced mortality; (2) the role of species interactions in the regulation of population density has been a recurring theme throughout the text; (3) R_o, first presented in life tables (see Chapter 18) is used as a determinant of whether a parasite population will increase; (4) the case of myxomatosis in European rabbits is an excellent example of a directional selection (see Chapter 21 - in fact, this is an example of two directional selections occurring simultaneously: reduced virulence in the myxomatosis virus and increased resistance in the rabbits).

The question of population regulation should be discussed thoroughly in the classroom, since it has relevance to human populations. Before modern medicine, epidemics and increased human mortality were common place; even today, such events occur in underdeveloped countries. In a broader context, you should emphasize the fact that, for regulation to take place, parasites must increase host mortality, decrease reproduction and must be maintained in host reservoirs during times of low host population density. Typically, regulation most likely would occur when parasites are of only moderate pathogenicity (parasites with severe pathogenicity presumably would not survive long enough for long-term regulation of a host population). In addition to any direct effects on host fitness, parasitism may act indirectly. For example, a heavy parasite load indirectly may influence the outcome of an interaction between the host and another species, e.g., lowered competitive ability, failure to escape predation, and increased susceptibility to a secondary infection because of parasite-induced stress.

In an evolutionary context, parasites and hosts surely must exert some selective pressure on one another. Parasites have evolved mechanisms to invade and escape hosts while at the same time overcoming defensive responses by the host. On the other hand, parasitic infestation presumably exerts a selection pressure on hosts to improve their defenses. One evolutionary question explored by the textbook relates to the virulence manifested by a parasite. Generally, low parasite virulence has been considered adaptive; however, the text describes a scenario in which high virulence also may be advantageous.

Chapter 26 concludes with a discussion of brood parasitism and kleptoparasitism, which are both forms of social parasitism. These interactions differ from the previous examples of parasitism in that they are not symbiotic. Brood parasitism can be subdivided into facultative and obligatory, as well as intraspecific and interspecific. The advantage of brood parasitism isn't readily obvious, but it is likely to be an evolutionary response to the need to reduce reproductive-related energy expenditures in habitats of poor quality where nesting resources are inadequate. Brood parasitism may or may not reduce the fitness of the host (see text). Kleptoparasitism is essentially the theft of food from a host and is a characteristic of many bird species. The obvious gain is energy, but this must be balanced against the quality of the food stolen and the energetic expenditure associated with all facets of theft, which ranges from time spent observing hosts to actual encounters with hosts.

TOPICS FOR DISCUSSION

1. Should the diversity of parasites be related to the diversity of host species? Should there be a relationship between host distributions and parasite distributions? Why or why not?

2. What are the similarities between herbivores and parasites? What are some differences?

3. Refer to review question eight in the textbook.

LECTURE OUTLINE CONTINUED:

I. Introduction to Parasitism

    A. Prevalence
    B. Hosts: Living Habitat
    C. Types of Parasitism
        1. Symbiotic Parasitism
        2. Social Parasitism
            a. Brood Parasitism
            b. Kleptoparasitism

II. Symbiotic Parasitism

    A. Modes and Dynamics of Transmission
    B. Effects on Hosts
    C. Evolutionary Considerations
    D. Population Responses
        1. Models of Parasitism
        2. Density and Dispersion
            a. Regulation of Host Density
            b. Parasite Population Dynamics

RESOURCE MATERIALS

In addition to a wide variety of parasitology textbooks that exist, a particularly useful reference for ecologists is by Esch and Fernandez (1993). This is an easily read book dealing with a wide variety of relevant topics, such as evolutionary aspects of parasitism, biogeography of parasitism, generalist versus specialist parasites, density-dependent influences on parasite populations, etc. For more general descriptions and examples, consider books by the following: Smyth (1994) for an introduction to animal parasitology, Blakeman and Williamson (1994) for plant diseases and Bogtish (1995) for aspects of human parasitology. For more information about parasite-host models and the parasite population dynamics, instructors also may want to augment the textbook's citations to work by R.M. Anderson and R.M. May with other papers by these individuals (May, 1981; Anderson, 1982). There are also many periodicals dedicated to studies of parasitology; some of the more prominent journals are: *Advances in Parasitology*, *Parasitology*, *Journal of Parasitology*, and *Parasitology Today*.

Anderson, R.M. 1982. "Epidemiology." In: *Modern Parasitology* (F.E.G. Cox, editor, pp. 204-251). Blackwell Scientific Publications, Oxford.

Bogtish, B.J. 1990. *Human Parasitology*. Saunders College Publishing, Philadelphia. PA.

Blakeman, B.P. and B. Williamson. 1994. *Ecology of Plant Pathogens*. CAB International, Wallingforg, England.

May, R. M. 1981. "Models for single populations." In: *Theoretical Ecology: Principles and Applications*, 2nd Edition (R.M. May, editor). Blackwell Scientific Publications, Oxford.

Esch, G.W. and J.C. Fernandez. 1993. *A Functional Biology of Parasitism - Ecological and Evolutionary Implications*. Chapman and Hall, New York.

Smyth, J.D. 1994. *Introduction to Animal Parasitology* (3rd Edition). Cambridge University Press, Cambridge.

CHAPTER 27

MUTUALISM

CHAPTER SUMMARY AND ORGANIZATION

Much confusion exists about mutualism, which has never received as much theoretical attention as competition and predation. Mutualism, however, may be one of the most important interactions. Unfortunately, much of what we know about mutualism is anecdotal; however, intriguing, highly-coevolved mutualisms are well-known. The ubiquitous nature of mutualisms cannot be overemphasized to students. Higher plants and animals may be examples of highly evolved, complex mutualisms. For instance, mitochondria and chloroplasts have prokaryotic DNA (food for thought can be found in Thomas, 1974, and Margulis, 1981 - see **Resource Materials**). The degree of mutualistic interactions involving prokaryotes and human function can be used to illustrate the prevalence of mutualism. In humans, prokaryotic cells outnumber eukaryotic cells by ten to one. In the mouth alone, over 100 species of prokaryotes can be found with more than 1,000 additional species also occurring in the gut. These organisms degrade saliva and mucus, secrete vitamins, as well as affect peristalsis, the immune system, and the size of the colon.

Chapter 27 begins with a consideration of coevolution. The criterion of coevolution dictates that two populations exert selective pressures on one another through evolutionary time. Coevolution typically is described as a game of adaptation and counter-adaptation involving taxa that impose selective pressures on one another. Many mutualisms are undoubtedly highly coevolved, but caution must be exercised not to extend the idea of coevolution to all mutualisms. In some instances, species may use traits evolved in other settings to invade new habitats and interact with newly encountered taxa; in other words, some species come into a new environmental setting preadapted for mutualistic interactions with appropriate taxa. The text also introduces "diffuse coevolution," in which selective pressures are exerted among groups or classes of species. In fact, general mutualistic relationships among several taxa are more common than specific interactions between pairs of species.

Mutualisms can be subdivided in many ways: nonsymbiotic versus symbiotic, obligate versus facultative, or direct versus indirect. The text integrates these types by first comparing obligate symbiotic and obligate nonsymbiotic mutualisms, after which facultative and defensive mutualisms are discussed. Of these, facultative mutualisms are the most common. To this point in the chapter, the types and examples of mutualism are "direct" forms in which species directly interact with one another. The text concludes the discussion of types of mutualism with a presentation of indirect mutualisms. The discussion of indirect mutualism can be described in an easily remembered way: friend's friends and enemies' enemies (see Boucher et al., 1982). The reasons for these names are readily apparent. If two species are separately mutualistic with a third species, then they can indirectly benefit one another through their positive effect on the third species, i.e., friend's friends. On the other hand, two species may benefit one another by having a common competitor that they both suppress, i.e., enemies' enemies. The work by Boucher et al. (1982 - see **Resource Materials**) can be very useful in augmenting classroom discussion of types and examples of mutualism. One key point of this paper is that mutualisms are not easily subdivided along the lines of symbiotic and nonsymbiotic interactions.

A thorough consideration of the reasons for being in a mutualistic relationship is as important as emphasizing the types of mutualism Most of the reasons are intuitively obvious and sections of Chapter 27 frame the basic possibilities. Reasons for the occurrence of mutualism include:

1. Enhanced nutrition and energy gain (e.g., breakdown of compounds for digestion by one partner, supply of growth factors or nutrients, attraction of prey items);

2. Protection (thoroughly discussed in the textbook section regarding defensive mutualisms);

3. Transport and dispersal of genetic information (thoroughly discussed in the textbook in the sections regarding pollination and seed dispersal);

4. Communication/mate attraction (evidence exists in oceanic profundal fish species that bioluminescent bacteria may aid in communication at depths beyond the penetration of light).

One of the most obvious questions to ask about mutualism is, "How do mutualisms arise?" Most direct mutualisms probably have evolved from a situation in which one taxon was exploited by another taxon to a situation in which both taxa became reciprocally exploitative. For example, some mutualisms may have evolved from parasitism (host: selection for resistance, tolerance and eventual use; parasite: selection for reduced virulence). Similarly, some mutualisms may have evolved from commensalism, if the commensal eventually provided some benefit that increased the host's chance of survival. Mutualisms involving gut microbes, prokaryotic organelles and endosymbiotic bacteria may have evolved from ingestion. In contrast to these possibilities, indirect mutualisms probably involved diffuse coevolution among a number of species that were not symbiotic with one another. This is very likely true for pollinators and seed dispersers.

A consideration of the effects of mutualism on population-level phenomena is the last point to be addressed in Chapter 27. In general, it is very difficult to demonstrate population effects derived from mutualisms. The text describes two examples of mutualisms (yucca plants - yucca moths and ants - acacia trees) in which reciprocal effects on population size seem evident. The text also includes an introduction to models of mutualism that are essentially a modification of the Lotka-Volterra equations for competition. As explained by the text, these models are overly simplistic and unrealistically lead to unlimited population growth. More complex models involve a check on population growth and interactions with other species.

TOPICS FOR DISCUSSION

1.   How is the diversity of vascular plants related to the variety and availability of pollinators?

2.   It has been hypothesized that mutualisms allow for survival in marginal habitats. What are some weaknesses and strengths of this argument? To what types of mutualism might this generalization be more applicable?

3.   How important is the mutualism between some plants and nitrogen-fixing bacteria to the nitrogen cycle and to the nitrogen needs of herbivores and carnivores? In other words, is this an example of a mutualism that has long-ranging effects well beyond the direct interaction between two taxa? Why or why not?

4.   How does the life cycle of most mutualists compare to the life cycles of parasites? Are they more or less complex? What are some possible reasons for differences? How do mutualism and parasitism affect the size of organismal ranges? In other words, do mutualisms tend to expand ranges or compress ranges relative to parasitic interactions?

5. Why is the term "mutualism" misleading in regard to the real nature of the interaction?

6. Refer to review questions 6-8 in the textbook.

LECTURE OUTLINE

I. Introduction

    A. Importance/Frequency of Mutualisms
    B. Historical Development of Terminology (see Boucher et al., 1982)
    C. Mutual Exploitation Rather Than Cooperation

II. Types of Mutualism

    A. Obligate versus Facultative
    B. Symbiotic versus Nonsymbiotic
    C. Direct versus Indirect
    D. Monophilic versus Polyphilic

III. Reasons for the Occurrence of Mutualistic Relationships

    A. Enhanced Nutrition and Energy Gain
        1. Breakdown of Compounds for Digestion
        2. Supply of Growth Factors and Nutrients
        3. Supply of Energy
    B. Protection
        1. From Other Taxa (Predators and Parasites)
        2. From Unsuitable Environmental Conditions
    C. Transport
        1. Of Pollen and Propagules
        2. Of Individuals
    D. Communication/Mate Attraction (Marine Profundal Bioluminescence)

IV. Evolution of Mutualism

    A. Introduction to Coevolution
    B. Possible Evolutionary Mechanisms for Mutualism
        1. Parasitism
        2. Commensalism
        3. Ingestion

V. Models and Effects of Mutualisms on Populations

RESOURCE MATERIALS

One particularly useful reference cited in the text is by Boucher et al. (1982); this review paper considers the origin of terminology, forms of mutualism and the evolution of mutualism. Bronstein (1995) also has written a recent and excellent review of mutualism. Issue no. 1 in volume 42 of *Bioscience* focuses on the phylogeny of plant-animal interactions and is an excellent source for information and ideas. Additional information about the evolution of mutualism can be found in the works by Howe (1984) and Sapp (1994). Sapp (1994) considers many topics, but chapter two explores the meanings of mutualism. A number of books about plant-animal interactions have chapters concerning mutualism, e.g., chapters two, three and six in Abrahamson (1989) deal with pollination biology, dispersal agents and ant-plant interactions, respectively.

Harborne (1988) also provides a chapter on the biochemistry of pollination. Particularly interesting examples of mutualism include the interactions between figs and wasps (Wiebes, 1979, cited below and in the textbook) and between ants and caterpillars (DeVries, 1992).

Abrahamson, W.G. (editor). 1989. *Plant-Animal Interactions*. McGraw-Hill Book Company, New York.

Boucher, D.H., S. James, and K.H. Keeler. 1982. "The Ecology of Mutualism." *Annual Review of Ecology and Systematics* 13:315-47.

Bronstein, J. 1994. "Our current understanding of mutualism." *Quarterly Review of Biology* 69:31-51.

DeVries, P.J. 1992. "Singing caterpillars, ants and symbiosis." *Scientific American*. 267(4):76-82.

Howe, H.F. 1984. "Constraints on the evolution of mutualism." *American Naturalist* 123:764-777.

Harborne, J.B. 1988. *Introduction to Ecological Biochemistry*. Academic Press, New York.

Margulis, L. 1981. *Symbiosis in Cell Evolution*. Freeman Publishers, San Francisco, CA.

Sapp, J. 1994. *Evolution by Association*. Oxford University Press, Inc., Oxford.

Thomas, L. 1974. *Lives of the Cell*. Bantam Books, Inc., New York.

Wiebes, J.T. 1979. "Coevolution of figs and their insect pollinators." *Annual Review of Ecology and Systematics* 10:1-12.

CHAPTER 28

COMMUNITY STRUCTURE

CHAPTER SUMMARY AND ORGANIZATION

Chapter 28 begins a three-chapter sequence that describes specific attributes of communities. This is a lengthy chapter dealing with many concepts, most of which build on ideas previously presented in the text. Most likely, the information in this chapter will take several lectures to complete and can be combined in various ways. The discussion of topics in Chapter 28 as presented below does not entirely follow the textbook's organization. Rather, vertical structure and classification are considered simultaneously, as are the questions of horizontal variation and community boundaries (ecotones). Other possibilities for departure from the text exist. For example, principles of island biogeography as related to maintenance of equilibrium species number easily could be considered in a discussion of species richness/diversity.

Any definition of a community is necessarily vague. The relevant features of a community require that an assemblage of species coexist together in a habitat and that they interact. Clearly, there are no size limits on communities. One can speak of the epiphytic community on a leaf, the bird community of a forest, or even the forest community. Delineation of a community is very much a function of personal perspective with no absolute criteria. In many instances, however, communities do have obvious boundaries, which often intergrade into other communities. Community edges and areas of community overlap (ecotones) are discussed in the third section of Chapter 28. Typically, ecotones are composed of species from each community type as well as species unique to the edge itself; consequently, species richness is often higher along community edges.

Historically, the nature of communities, plant communities in particular, has been intensely debated. Are communities more or less random collections of species with similar habitat requirements? Or, are communities complex associations of species that are highly dependent on one another? Are communities discontinuous or is there broad overlap? The answers to these questions were once strongly related to two competing perceptions of the community that originated between 1916 and 1926: the organismic concept versus the individualistic concept. The nature of this conflict was first introduced in Chapter 1 of the textbook and is further expanded upon in Chapter 28; in addition, the significance of these opposing views to the development of successional theory is presented in Chapter 30. Most contemporary ecologists subscribe to the individualistic view that species are distributed primarily according to habitat requirements while recognizing that strong associations between some species can occur. There is no example of a community in which all species are bound together by obligatory relationships. Gradient analysis and ordination procedures were the driving force in the 1950s-1960s that clarified the nature of plant communities. Activities/analyses similar to these early studies make good laboratory exercises for the study of plant communities. Rolling topography sloping down to a river often provides an excellent habitat for assessing species distributions along a moisture gradient.

The opening and concluding sections of Chapter 28 discuss the physical structure of communities and community classification. Instructors may want to combine these sections in a lecture, since classification of most communities is in large part based on the structure of the plant community. The fact that most communities are classified on the basis of plants should not be surprising to students. Plants represent the most stable living component of the community. Because they lack mobility, plants must cope with the prevailing environment; hence, they often make excellent indicators. Given their autotrophic capabilities, plants set the stage for energy flow through food chains/webs. Furthermore, as a generalization, the distributional patterns of animals often are correlated with plant communities. Typically, community classification schemes include a consideration of community "physiognomy." Physiognomy is essentially the appearance of vegetation; students can think of physiognomy as "what they see" when they are looking at a community. Students should immediately "see" two things: (1) various plant life forms and (2) some degree of vertical structure. In combination, life form and vertical structure represent community physiognomy. A plant's life form can be considered from any number of viewpoints, but basically can be summed up in three words: "mode of existence." In other words, life form is a combination of structural and physiological traits that allow an organism (plant in this case) to exist within a particular environment. When students look at a plant community, they may see herbs, shrubs, trees or vines in various proportions, depending on the community. Or, they may see leaves that are needle-shaped and evergreen, or broad and deciduous, or the leaves may even be absent. These are a few representations of various morphological life form criteria, but life form criteria with a physiological basis must not be overlooked. For example, C_3, C_4 and CAM photosynthetic systems represent a life form sequence adapted to deal with increasing levels of aridity. Vertical structure merely refers to the various strata that may exist in a plant community. For example, a forest may have four layers (floor/herb layer, shrub layer, subcanopy and canopy). A desert (e.g., Mohave Desert), may have only two layers. Grasslands essentially have a grass layer, a litter layer and a well-developed fibrous root zone. In combination, plant life form and vertical structure represent visual images of specific community types. In other words, communities can be separated or combined on the basis of the uniformity of their physiognomy. Community classification also may include a consideration of species composition and prevailing environmental regimes. Although communities can be classified into groups, students should not be lulled into thinking that communities are completely homogeneous.

In its discussion of vertical structure, the text includes a description of the Raunkiaer life form spectrum, which is based on the position of a plant's perennating bud above the ground. Raunkiaer life forms can be used as one general descriptor of a community, but caution should be exercised when using these life forms as a basis for community classification. Although broad generalizations are possible and biologically meaningful, overlap does occur among the Raunkiaer life form spectra of various communities, with no one community represented by a specific life form and no life form occurring in only one community. About the only real generalization that can be made is that therophytes are very prevalent in fire-dominated habitats.

In contrast to vertical structure, horizontal structure is far less predictable. Communities tend to consist of patches existing at a variety of spatial scales. The pattern of patches is indicative of the interaction of abiotic and biotic factors, which affect organismal distribution.

The second major section of Chapter 28 focuses on species dominance, diversity and abundance in communities, each of which is a component of the biotic structure of a community. Ideas about diversity, dominance and relative measures of abundance are related and can be unified for lectures. Species diversity usually is determined by some index that combines species richness with a measure of species evenness or equitability. Communities with low evenness are dominated numerically by a few species. The major portion of this section is dedicated to a discussion of species diversity and the major factors affecting diversity. The text presents the Shannon (or Shannon-Wiener) index of diversity, which determines the uncertainty associated with two individuals, drawn at random from a community, belonging to the same species. Obviously, such uncertainty will be the greatest when species richness and species evenness are high. Simpson's Index and Brillouin's index are two other common measures of diversity not discussed by the text (see Krebs, 1989 - **Resource Materials**). No general agreement exists about how to characterize community diversity and, in some instances, the existing indices can be misleading (see Hurlbert, 1971; Pielou, 1975; Connell, 1978). However, diversity indices can be particularly useful in an applied context and can be used to document habitat degradation. There are many determinants of diversity; the text presents hypotheses regarding evolutionary time, ecological time, spatial heterogeneity, climatic stability, climatic predictability, energy, productivity, competition, predation, dynamic equilibrium and stability-time. Valuable laboratory exercises can be devised to test some of these hypotheses (students can collect information from the field on species abundances in contrasting habitats and use one of the diversity indices; plants and invertebrates are perhaps the easiest organisms with which to work). Probably at least one lecture will need to be devoted to a discussion of diversity. This lecture also can be supplemented with contemporary losses of biodiversity, both in terms of species losses and losses of genetic information.

Chapter 28 also presents an overview of traditional island biogeography theory. Theories about species immigration, extinction, and turnover on small versus large islands and on islands near to and far from a mainland are presented. While these models are overly simplistic, they do provide a basis for considering the island-like nature of many habitats (e.g., lakes, mountain tops, wildlife preserves, etc.) and some of the consequences of habitat fragmentation. In the latter context, various habitat management strategies are considered in detail by the textbook. Arguments persist about minimum habitat size, whether to manage habitats for maximum species diversity, or to manage for maintenance of specific species. Smaller habitats often have a higher species diversity than larger habitats, but some species are poorly represented (e.g., interior species and area-sensitive species). Questions of minimum habitat size can be answered only on a species-specific basis. The text also discusses the value of corridors for providing dispersal routes for species between habitat islands.

Another important question in community ecology relates to the role of organismal interactions in affecting community structure. By this point in an ecology course, students should be thoroughly conversant with the topics of competition, predation, parasitism and mutualism (Chapters 22 - 27). The textbook approaches these interactions in Chapter 28 from the standpoint of whether they affect community structure/organization. The meaning of "community structure" is historically ambiguous, but, as defined by the text, structure relates primarily to facets of resource utilization, niche segregation, species diversity/abundance, organization of food webs, etc. Although no clear-cut generalizations are possible about the role of these interactions in shaping the structure of all communities, the text does present numerous examples in which a particular interaction plays some significant role. Historically, competition has been assumed to be the dominant interaction, but examples of competition are often less definitive than for predation and parasitism. Mutualism has not been studied enough, but could well be one of the most important determinants of community structure. As these ideas are discussed, you should require students to draw upon their knowledge of organismal interactions to predict effects on community structure.

TOPICS FOR DISCUSSION

1. What are some examples of habitat islands in suburban settings? Consider parks, golf courses, etc. How prevalent are habitat corridors between these areas? Design laboratory activities to compare the richness and/or diversity of various taxa in these areas as a function of the habitat size.

2. Habitat fragmentation has been more responsible for the dramatic increase in whitetail deer density in the midwestern U.S. than removal of predators. Why? What does this tell us about the habitat requirements of deer?

3. Refer to review questions 14-16 in the textbook.

LECTURE OUTLINE

Information in the following outline will require <u>at</u> <u>least</u> three one-hour lectures. Items I, II and III can be covered in one lecture. Items IV and V could perhaps be covered in one lecture, but two lectures seem more probable. Items VI and VII shouldn't require more than one lecture. Of course, this is only one of many possible organizational schemes. The text presents a substantially different organizational scheme with the same content.

I. The Nature of the Community

A. Definition of Community
B. Organismal versus Individualistic Community Concepts

II. Classification of Communities

A. Classification Based on Plant Assemblages
B. Physiognomy
    1. Vertical Architecture (Layers)
    2. Life Forms
        a. Examples of Life Form Criteria
        b. Raunkiaer Life Forms
C. Species Composition - The Association (not discussed in text)

III. Horizontal Structure

A. Heterogeneity: Patchy Species Distributions
B. Edges and Ecotones

IV. Species Diversity

A. Species Dominance and Relative Measures of Abundance/Importance
B. Measures/Indices of Species Diversity
    1. Species Richness and Evenness
    2. Shannon-Wiener Index
    3. Simpson's Index (not discussed in text)
C. General Global Patterns of Diversity
D. Determinants of Diversity: Hypotheses
    1. Evolutionary Time Hypothesis
    2. Ecological Time Hypothesis
    3. Spatial Heterogeneity Hypothesis
    4. Climatic Stability Hypothesis
    5. Climatic Predictability Hypothesis
    6. Energy Hypothesis
    7. Stability-Time Hypothesis
    8. Productivity Hypothesis
    9. Competition Hypothesis
    10. Predation Hypothesis (Cropping Principle)
    11. Dynamic Equilibrium Hypothesis

V. Species Abundance

A. Random Niche Model
B. Log-Normal Model
C. Niche Preemption Model

LECTURE OUTLINE CONTINUED:

VI. Biotic Structure and Population Interactions

    A. Role of Competition
    B. Role of Predation
    C. Role of Parasitism
    D. Role of Mutualism

VII. Island Communities

    A. Principles of Island Biogeography
        1. Extinction versus Immigration
        2. Equilibrium Species Number
        3. Large versus Small Islands
        4. Near versus Far Islands (with respect of proximity to mainland areas)
    B. Examples of "Islands" in Terrestrial Settings
        1. Bogs, Lakes, Wetlands
        2. Mountain Tops
        3. Sand Dunes
        4. Man-Made Islands: Parks, Wildlife Refuges, Green Spaces, Golf Courses, etc.
    C. Habitat Fragmentation
        1. Consequences
        2. Debate About Size of Preserves
        3. Value of Habitat Corridors

RESOURCE MATERIALS

Since a large portion of Chapter 28 deals with species abundance and diversity, many of the references below augment this discussion. The citations range from general (e.g., Huston, 1979) to more detailed discussions of diversity in particular ecosystems (e.g., Connell, 1978; Denslow, 1987). Questions of ideal levels of species richness are pondered by Baskin (1994). Wilson's book (1992) is directed to non-scientists, but provides a good overview to the problems of understanding diversity and the contemporary decline in biodiversity. Giller (1984) includes information about species diversity as well as community structure and island biogeography.

Chapter eight in the book by Barbour et al. (1987) considers community concepts (particularly the continuum concept) and community attributes (including a discussion of physiognomy). Austin (1985) also examines the continuum concept, emphasizing the similarities between the continuum concept in plant ecology and the niche theory from animal ecology. Diamond and Case (1986) provide a collection of papers regarding communities; in particular, see sections five and six for information about the factors which structure communities. Another collection of papers (Strong et al., 1984) also contains useful information; see chapters relating to island biogeography and community structure. Risser (1995) provides an up-to-date treatment of the numerous roles of ecotones.

Austin, M.P. 1985. "Continuum concept, ordination methods and niche theory." *Annual Review of Ecology and Systematics* 16:39-61.

Barbour, M.G., J.H. Burk, and W.D. Pitts. 1987. *Terrestrial Plant Ecology*. The Benjamin/Cummings Publishing Company, Inc., Menlo Park, CA.

Baskin, Y. 1994. "Ecosystem function of biodiversity." *Bioscience* 44:657-660.

Connell, J.H. 1978. "Diversity in tropical rain forests and coral reefs." *Science* 199:1302-1310.

Denslow, J.S. 1987. "Tropical rainforest gaps and tree species diversity." *Annual Review of Ecology and Systematics* 18:431-451.

Diamond, J. and T.J. Case. 1986. *Community Ecology*. Harper and Row Publishers, New York.

Giller, P.S. 1984. *Community Structure and the Niche*. Chapman and Hall, New York.

Hurlbert, S.H. 1971. "The nonconcept of species diversity: A critique and alternative parameters." *Ecology* 42:577-586.

Huston M. 1979. "A general hypothesis of species diversity." *American Naturalist* 113:81-101.

Krebs, C.J. 1989. *Ecological Methodology*. Harper and Row Publishers, New York.

Peet, R.K. 1974. "The measurement of species diversity." *Annual Review of Ecology and Systematics* 5:285-307.

P.G. Risser. 1995. "The status of the science examining ecotones." *Bioscience* 45:318-325.

Schafer, C.L. 1995. "Values and shortcomings of small reserves." *Bioscience* 45:80-88.

Strong, D.R., D. Simberloff, L.G. Abele, and A.B. Thistle. 1982. *Ecological Communities: Conceptual Issues and the Evidence*. Princeton University Press, Princeton, NJ.

Wilson, E.O. 1992. *The Diversity of Life*. Harvard University Press, Cambridge, MA.

CHAPTER 29

DISTURBANCE

CHAPTER SUMMARY AND ORGANIZATION

Chapter 29 defines disturbance as "any physical force, such as fire, wind, flood, extremely cold temperature and epidemic, that damages natural systems and results in the mortality of organisms or loss of biomass." The discussion of disturbance and its effects form a bridge between the material presented about species diversity (Chapter 28) and the material presented about succession (Chapter 30). Disturbance creates colonization sites, thereby increasing the abundance of opportunistic species and diversity while simultaneously initiating secondary successions. Disturbances can be characterized on the basis of intensity, frequency and scale. Intensity is a measure of the magnitude of the physical force of the disturbance, usually expressed in terms of the removal or mortality of individuals, species and/or biomass. Frequency is the rate of disturbance (# disturbances/time). Scale is rather abstract, but refers to the size of the disturbance. Scale of disturbance must be considered in the context of the scale of the community being affected. For example, a single treefall may be a significant local disturbance, but it is probably inconsequential to a forest in its entirety. Each of these characteristics is discussed in the first section of the chapter.

The second section of Chapter 29 details some of the most probable kinds of disturbances. This section is important for making students aware that most disturbances are natural and that all communities (and ecosystems) experience disturbances on some time scale. Most students initially will think only of man-made disturbances. This section opens with a detailed presentation of disturbances by fire. Surface, crown and ground fires are described. Of these, the most destructive is the ground fire, which consumes organic matter down to inorganic substrate. Unable to directly escape fire, plants have evolved various adaptations for withstanding periodic fires. These adaptations are described in some detail, but can generally be considered in the context of "seeders, sprouters and tolerators." Some plants protect their seeds to the extent that fire might even be required for germination. Other plants sprout new shoots from protected buds and underground structures. Tolerators have thick bark or other adaptations to withstand intense heat.

Other sources of disturbance that are described in Chapter 29 result from wind, moving water (erosion and deposition), drought and animal activity (grazing in particular). Three forms of man-made impacts also are discussed: timber harvesting, cultivation and surface mining. Obviously, the North American ecological landscape has been altered dramatically by human practices for over two centuries. European natural communities were altered centuries earlier and almost every student will be familiar with contemporary tropical deforestation and slash-and-burn agriculture. In most cases, the absence of any reliable data prior to man-made impacts prevents us from knowing the true level of devastation that may have occurred or may be occurring. In an era of heightened, yet insufficient, environmental awareness in some developed countries, ecosystem reclamation and restoration are popular environmental themes with great sums of money being spent on these activities. While such activities are laudatory, the reality is that an ecosystem seldom can be restored to its original condition; this is particularly true for complex systems such as forests and oligotrophic lakes. Students should not think that man's activity can be undone despite our best intentions. In the case of habitat/ecosystem protection, the old adage, "An ounce of prevention is worth a pound of cure," is apropos.

The text also considers the specific effects of disturbance on nutrient cycling. Many disturbances, particularly those associated with biomass removal, result in a loss of nutrients from the ecosystem. The nutrient levels decline because of direct removal and because of increased erosion and leaching. In other situations, a disturbance may result in a flush of nutrients, (e.g., from a fire); but, if these nutrients are not sequestered through physical or biological means, they may still be lost through erosion and leaching. As the text points out, retention of nutrients after a disturbance is the key to maintaining balanced nutrient cycles.

The extent to which animal species are affected by disturbance depends on the animal in question. Most of the text's discussion centers on fire disturbance. In some cases, an animal species may require fire for its existence, and the text discusses the fire-dependent Kirtland's warbler as an example. Most mobile organisms, such as birds and mammals, and most ground-dwelling organisms experience only minor negative effect from fire. In other instances, animal species may suffer, at least in the short-run, because of habitat destruction and/or loss of resources. In a broader context, probably the most devastating effects on animals come from whole-sale destruction of habitats by humans.

The last section of Chapter 29, "Community Stability" (or equilibrium), is particularly important and describes the concepts of community resistance and resilience. Resistance represents a system's capacity to resist variation. Resilience is the speed at which a system returns to its original condition prior to the disturbance. Stability, resistance and resilience are important ecosystem attributes that presumably operate when a community is at equilibrium. Most communities, however, continually experience disturbances of one sort or another and, therefore, may be in a continuous state of nonequilibrium. Species diversity may be higher in a nonequilibrium community provided that environmental disturbances are not too frequent or too far apart (intermediate disturbance hypothesis - see Chapter 28).

TOPICS FOR DISCUSSION

1. What are the effects of tree removal (by fire or clear-cutting) on forest stream communities? Consider parameters such as water temperature, turbidity, nutrient concentrations, sedimentation, etc. (see work performed in Hubbard Brook Experimental Forest - Bormann and Likens, 1994 - **Resource Materials**).

2. Disturbance is usually perceived in a negative context, but what kinds of disturbances function as renewing an ecosystem or at least some components of an ecosystem?

3. Human activity usually is associated with direct disturbance to ecosystems. What are some examples in which we have tried to suppress the frequency of various forms of disturbance? Is such suppression a form of disturbance in itself? What are the consequences of suppressing disturbance?

4. Why does so much energy have to be placed on weed control in cultivated fields? Consider the frequency of disturbance in such settings and the kind of habitat provided by cultivation. Does the creation of habitat for opportunistic species followed by intense efforts to control their abundance make good intuitive sense?

5. Refer to review questions 10-11 in the textbook.

LECTURE OUTLINE

I. Disturbance: General Considerations

    A. Intensity
    B. Frequency
    C. Scale

II. Causes of Disturbance

    A. Fire
        1. Fire Climate
        2. Fire-Adapted Ecosystems
        3. Types of Fires
            a. Surface Fire
            b. Crown Fire
            c. Ground Fire
        4. Positive Effects of Fire
        5. Negative Effects of Fire
    B. Wind
    C. Water Movement
        1. Flooding
        2. Corrasion
        3. Sediment Deposition
    D. Drought
    E. Animal Activity
    F. Examples of Human Disturbances
        1. Timber Harvesting
        2. Cultivation
        3. Livestock Grazing
        4. Dam Construction
        5. Mining

III. Effect of Disturbance On Communities

    A. Equilibrium Communities
        1. Resistance and Stability
        2. Resilience
    B. Nonequilibrium Communities and the Intermediate Disturbance Hypothesis

RESOURCE MATERIALS

Many general works on disturbance exist. Sousa (1984) provides an excellent, comprehensive review that places disturbance into the context of its ecological effects; much of this article deals with patch dynamics as related to disturbance. Other general monographs include those by Mooney and Godron (1983), Pickett and White (1985) and Goigel (1987). The ecological implications of some recent, major disturbances in the U.S. are interesting topics to include in the classroom; e.g., see Allen's article (1993) for insights into flooding along the Mississippi (and elsewhere), and refer to volume 44, issue 4 of *Bioscience* (1994) for a consideration of the effects of Hurricane Andrew on southeastern U.S. ecosystems. One common ecosystem disturbance is fire, and many articles and monographs have addressed fire ecology (one classic paper is by Daubenmire, 1968). Wright and Bailey (1982) offer an excellent book covering fire effects and vegetative response in grasslands, conifer forests and shrublands.

The text does not consider in any detail the importance of disturbance in aquatic ecosystems. Padisák et al. (1993) address the intermediate disturbance hypothesis to aquatic ecosystems. In particular, the role of disturbance in affecting phytoplankton diversity, seasonal successions and community structure is considered.

Chapter 29 discusses the consequences of human-related disturbances associated with timber harvest, cultivation and mining. Humans, however, produce other major disruptions in ecosystems. A particularly good book that considers some of the disturbances (and stresses) that result from human activity is by Freedman (1989). This book addresses the effects of forest harvesting, but also includes the potential consequences due to eutrophication and oil, acid, air and pesticide pollution; an interesting chapter on the effects of war on ecosystems concludes this book. Additionally, one might examine Volume 23 (1992) of the *Annual Review of Ecology and Systematics* for global changes related to human activity.

Allen, W.H. 1993. "The Great Flood of 1993." *Bioscience* 43:732-737.

Bormann, F.H. and G.E. Likens. 1994. *Pattern and Process in a Forested Ecosystem: Disturbance, Development, and the Steady State based on the Hubbard Brook Ecosystem Study*. Springer-Verlag, New York.

Daubenmire, R. 1968. "Ecology of Fire in Grasslands." *Advances in Ecological Research* 5:209-266.

Freedman, Bill. 1989. *Environmental Ecology - The Impacts of Pollution and Other Stresses on Ecosystem Structure and Function*. Academic Press, Inc., New York.

Goigel, M. (Editor). 1987. *Landscape Heterogeneity and Disturbance*. Springer-Verlag, New York.

Mooney, H.A. and M. Godron. (Editors). 1983. *Disturbance and Ecosystems - Components of Response*. Springer-Verlag Inc., New York.

Padisák, J., C.S. Reynolds, and U. Sommer. 1993. *Intermediate Disturbance Hypothesis in Phytoplankton Ecology*. Kluwer Academic Publishers, Dordrecht, The Netherlands.

Pickett, S.T.A. and P.S. White (editors). 1985. *The Ecology of Natural Disturbance and Patch Dynamics*. Academic Press, New York.

Sousa, W. "The role of disturbance in natural communities." *Annual Review of Ecology and Systematics* 15:353-391.

Wright, H.A. and A.W. Bailey. 1982. *Fire Ecology*. John Wiley and Sons, Inc., New York.

CHAPTER 30

SUCCESSION

CHAPTER SUMMARY AND ORGANIZATION

Long considered a unifying ecological principle, the concept of succession and its mechanisms remained unchallenged for many years. Introduced by Henry Cowles in 1899 and expanded upon by Frederick Clements, succession was viewed as "a continuous, unidirectional, sequential change in the species composition of natural communities" (see text) and was intimately associated with Clements's organismic view of the community. Clements viewed succession as a predictable, inevitable process driven by the action of plants on their environment (facilitation) that concluded with a stable endpoint determined by the prevailing climate. Amazingly, this is the view that many life-science students still hear in introductory biology courses and even in general ecology courses. Although some of Clements's six-step hypothesis of succession (initiation, immigration, establishment, competition, site modification and stabilization) can be found in many modern approaches to succession, new studies have revealed further insight into questions about mechanisms of vegetation change, the nature of the climax (if indeed it exists) and whether successional changes are truly predictable and directional. These new insights have created productive debates among ecologists, particularly between reductionist and holistic approaches, as well as between advocates of the organismic versus individualistic views of communities. Chapter 30 provides a good introduction to the contemporary understanding of succession while providing a strong historical perspective and emphasis on commonalities between the recent studies and older work.

After an initial overview to successional terminology (pioneer species, climax, autogenic, allogenic, etc.), the text describes classic examples of primary terrestrial succession (sand dunes, and floodplain deposits) and secondary terrestrial succession (abandoned fields). The information about disturbances in Chapter 29 also can be used to emphasize how many secondary successions are initiated. The text also describes examples of succession in aquatic habitats, ponds and intertidal areas. These examples are all directional. If one component of a lecture about successional theory encompasses a consideration of types of succession, consider including the examples of cyclic succession as a contrast to directional succession. The text includes a discussion of cyclic succession under the topic of the climax (fourth section of Chapter 30). The text refrains from using some conventional terminology, such as xerarch and hydrarch succession, perhaps because the terms may conjure up outdated ideas about succession; nonetheless, many students probably will have heard these terms previously and will inquire about them. Instructors also may want to consider presenting examples of fluctuations as a contrast to the examples of succession at this point in a lecture (fluctuations are presented in the sixth section of the chapter).

The third section of Chapter 30 describes successional models, and the nature of early controversies becomes obvious. This section opens with a discussion of Clements's original ideas, which were an extension of his organismal view of the community. This discussion is then followed by an explanation of Henry Allen Gleason's (Clements's main initial challenger) ideas about succession, which were consistent with his view of a community as a collection of individuals responding to specific environmental conditions. Clements's ideas also formed a basis for the holistic approach to succession, which was in direct contrast to the reductionist concept. Reductionists view succession as a product of population dynamics in comparison to the holistic viewpoint that emergent community properties exist and develop as a function of ecosystem organization. In 1977, Connell and Slayter (see text) presented three models that encompassed elements from both the holistic and reductionist schools of thought. One model, the facilitation model, is holistic and Clementsian and emphasizes autogenic habitat changes induced by the organisms themselves. The second model, the inhibition model, is essentially a reductionist approach with competition driving vegetational change. The third model, the tolerance model, is more involved and includes elements from both schools. The tolerance model suggests that early stages of succession are driven by competition, whereas later stages are dominated by species that can invade and tolerate lower resource regimes than earlier-existing species on the site; clearly, this model depends on alternative life history characteristics. The text does not discuss John Grime's model of succession (1979), which also deals with plant life history strategies. Grime's model considers plant succession to be a sequence of life history strategies with ruderals invading a site, followed by competitors and eventually stress tolerators (r,- C -and S -selection - see Chapter 20).

The mechanisms of succession often are debated and obviously are included in some of the above models (e.g., mechanistic roles of facilitation versus competition). Actual mechanisms causing plant change are probably a combination of many factors operating simultaneously. One model that integrates both specific characteristics of the ecosystem and competition is David Tilman's "resource ratio model" (1985, 1988 - see text). This hypothesis considers competitive effects in conjunction with gradients of specific limiting resources (primarily light and soil nutrients). Changing availability of and competition for these resources are the bases for plant change. A second model discussed by the text is based on a combination of factors that operate at the level of the individual, i.e., an individual-based model. Again, competition is a forcing function in this model, in which community change is driven by the interaction of specific plant life history/physiological traits and changing environmental regimes. Whether one considers populations or individuals, these models have the following points in common: numerous successional patterns are possible, competition is a key component of species replacement, and environmental conditions change because of the autogenic effect of plants on their habitat.

Ideas about the nature and existence of a successional climax as described in the fourth section of the chapter also have evolved since first presented. Clements first described the *monoclimax* theory (also termed the Clementsian climax and climatic climax). This is a rigid view in which a successional sequence has only one eventual climax, which is determined by the prevailing regional climate. While no one would argue that climate is a dominant force, there are problems, of course, with this approach because within a region of homogeneous climate one finds many plant communities that seem stable with little evidence of change. Consequently, the *polyclimax* theory was proposed in which a mosaic of different plant communities is attributed to various environmental phenomena other than climate, e.g., soil conditions, fire, and topography - edaphic, pyric and topo climaxes, respectively. Hence, a number of climaxes are recognized. This theory is essentially Clementsian and the climax is determined by a single environmental factor. The *pattern climax* theory considers the vegetative climax to be a product of the total environment, whereby numerous factors interact to determine species composition. Communities appear as mosaics of climax vegetation because of environmental gradients that overlap and intersect.

Some ecologists would argue that the idea of a stable, self-perpetuating climax is a conceptual myth rather than a biological reality. There are several arguments that favor this viewpoint. First, no community is stable forever; time and stability are merely matters of perspective (see time and direction in succession - eighth section of Chapter 30). Second, there are many known examples of cyclic succession that never achieve a stable climax. Third, a few studies over the last decade have indicated that communities perceived as climax are actually in a state of flux. This latter point is emphasized in the textbook with examples of studies in a Tennessee hardwoods forest and a Douglas fir forest of the Pacific Northwest. These studies do indicate that although species composition is unstable, overall physiognomy does not vary appreciably, which results in a *shifting-mosaic steady state*.

For many years, hypotheses about changes in ecosystem attributes during directional succession had achieved the level of ecological dogma. These attributes and the expected trends are outlined in Table 30.2. An important aspect of the table that <u>must</u> be emphasized in the classroom is whether or not to accept or reject these hypotheses. In general, the only hypotheses that can be accepted from Table 30.2 are that biomass, total organic matter, species richness and the ratio of biomass to ecosystem respiration increase. Conversely, animal net community production and the ratio of gross primary production to biomass both decrease. Some instructors may find it useful to combine information from Table 30.1 (life history characteristics of plants from early to late successional stages) with the appropriate information from Table 30.2 to provide a total portrayal of the kinds of changes that might occur during a successional sequence. One important aspect of vegetational/community change that also should be addressed is the change in animal communities. This topic is addressed under the heading of "succession and animal life" in the ninth section of Chapter 30. Students should be aware that the existence of many animal species are highly dependent on the presence of specific successional stages; the text presents such examples for the golden-winged warbler and the American woodcock. Consequently, the key to high animal diversity is the presence of a variety of successional stages across a landscape. This section concludes with a note that some animals also influence succession, particularly grazers.

The final two sections of Chapter 30 describe serial replacements associated with the decay of organic matter (degradative or heterotrophic succession) and geologic time periods (paleosuccession). The decay/degradation of organic matter involves a sequence of organismal invasions and replacements as the organic base is altered. The text describes Winston's (1956) study of degradative succession on acorns. Other examples that could be used are fungal succession on pine litter (Kendrick and Burges, 1962 - see **Resource Materials**) or aquatic plant material (Morrison et al., 1977). The point of emphasis for the section about paleosuccession is for students to understand that vegetational changes also occur along time scales that far exceed the lives of individuals, or even human cultures, for that matter. The text describes changes that occurred during and before the Pleistocene epoch. In particular, contrasts are drawn between the effect of advancing glaciers on North American and European vegetation.

In conclusion, coverage of the theory of succession must be designed not only to provide a thorough treatment of succession, but also to dispel many of the misperceptions that students will have about succession. The only introduction most students will have had of succession probably centered on a directional, facilitation model of succession culminating in a monoclimax. Clearly, this is an outmoded view of succession that must be placed into its proper context.

TOPICS FOR DISCUSSION

1. Is succession a logical, orderly process? Can the actual order of species occurrences be predicted accurately? Why or why not?

2. Consider the spatial scale of succession. Humans think in terms of what they see, but are there examples of microbiotic successions? Consider microcommunities of fungi, algae, bacteria, protozoans, nematodes, etc., on the surface of aquatic plant leaves.

3. Do successions have an endpoint? Why or why not?

4. Since succession involves vegetation change, why do we use the term succession. See Tansley (1935 -**Resource Materials**) for possible insights. Given our new information about succession, is Tansley's assertion that succession involves the recognition of a sequence of phases still valid?

LECTURE OUTLINE

I. Types of Vegetational Change

    A. Succession
- a. Overall Concept and Definition
- b. Types of Succession & Terminology
  - *i.*   primary succession
  - *ii.* secondary succession
  - *iii.* allogenic succession
  - *iv.* autogenic succession
  - *v.*   cyclic succession

    B. Fluctuations

II. Successional Models and Mechanisms

    A. Clements's Concept of Succession
    B. Gleason's Concept of Succession
    C. Holistic versus Reductionist Approach
    D. Connell-Slayter Models
- 1. Facilitation Model
- 2. Inhibition Model
- 3. Tolerance Model

    E. Grime's Model: r-, C- and S-selected strategies (not presented in text)
    F. Tilman's Resource Ratio Model
    G. Individual-Based Models
    H. Potential Role of Herbivores

III. The Climax Concept

    A. Clementsian Monoclimax Theory
    B. Tansley's Polyclimax Theory
    C. Whittaker's Pattern Climax Theory
    D. Shifting-Mosaic Steady State

IV. Hypothesized Trends in a Directional, Progressive Succession

    A. Changes in Life History Strategies (see Table 30.1)
    B. Changes in Ecosystem Attributes (see Table 30.2)

V. Successional Stages and Animal Life

VI. Other Successional Patterns

    A. Degradative Succession
    B. Paleosuccession

RESOURCE MATERIALS

Chapter 30 provides citations to some important papers in the primary literature, ranging from classic studies to contemporary evaluations of succession. Particularly useful references that instructors should read are the article by Bazzaz (1979) and the citations to D. Tilman's work. A variety of other monographs and studies also will prove useful. Gray et al. (1987) provides an excellent book, which begins with an overview to succession and includes detailed chapters about topics not covered in the textbook, including trends in herbivory during succession, genetic changes that may occur along a successional sequence, assembly rules during succession, the stability of communities, etc. Another particularly useful book is provided by Golley (1977); this work is a collection of seminal papers and will provide an excellent overview to the evolution of thought about succession. Matthews's book (1992) is valuable for its specific examples, many of which relate well to the textbook. West et al. (1981) discuss multiple forms of forest succession with examples from different localities. Miles (1979) offers a short work, but it contains a wealth of information about vegetation change.

Information about primary succession is abundant. A recent examination of primary succession in Glacier Bay concluded that many factors interact to determine successional patterns with plant life history traits dominating (Chapin et al., 1994). Interesting information about succession after the eruption of Mount St. Helens can be found in the article by Moral and Bliss (1993); this information also might be useful for background information to supplement lectures about the significance of disturbance (Chapter 29). Consult the work by Horn (1974) for information about secondary succession, particularly for a consideration of potential changes in diversity and stability during succession. For examples of heterotrophic and/or microbial successions, refer to the works by Morrison et al. (1977) and Kendrick and Burges (1962).

Bazzaz, F.A. 1979. "The physiological ecology of plant succession." *Annual Review of Ecology and Systematics* 10:351-371.

Burrows, C.J. 1990. *Processes of Vegetation Change*. Unwin Hyman Academic, New York.

Chapin, F.S., L.R. Walker, C.L. Fastie, and L.C. Sharman. 1994. "Mechanisms of primary succession following deglaciation at Glacier Bay, Alaska." *Ecological Monographs* 64:149-175.

Glenn-Lewin, D.C. and R.K. Peet. 1992. *Plant Succession - Theory and Prediction*. Chapman and Hall, New York.

Gray, A.J., M.J. Crawley, and P.J. Edwards. 1987. *Colonization, Succession and Stability*. Blackwell Scientific Publications, London.

Golley, F.B. 1977. *Ecological Succession*. Dowden, Hutchinson and Ross, Inc., Stroudsburg, PA.

Horn, H.S. 1974. "The ecology of secondary succession." *Annual Review of Ecology and Systematics* 5:25-37.

Kendrick, W.G. and A. Burges. 1962. "Biological aspects of the decay of *Pinus sylvestris* leaf litter." *Nova Hedwiga* 4:313-342.

Matthews, J.A. 1992. *The Ecology of Recently-Deglaciated Terrain: A Geoecological Approach to Glacier Forelands and Primary Succession*. Cambridge University Press, Inc., Cambridge.

Miles, J. 1979. *Vegetation Dynamics*. Chapman and Hall, London.

Moral, R. Del and L.C. Bliss. 1993. "Mechanisms of primary succession: insights resulting from the eruption of Mount St. Helens." *Advances in Ecological Research* 24:1-69.

Morrison, S.J., J.D. King, R.J. Bobbie, R.E. Bechtold, and D.C. White. 1977. "Evidence of microfloral succession on allochthonous plant litter in Apalachicola Bay, Florida, U.S.A." *Marine Biology* 41:229-240.

Tansley, A.G. 1935. "The use and abuse of vegetational concepts and terms." *Ecology* 6:284-307.

West, D.C., H.H. Shugart, and D.B. Botkin. 1981. *Forest Succession: Concepts and Application*. Springer-Verlag, New York.

Multimedia Resources

Insight Media, 2162 Broadway, New York, NY 10024. --- "**Succession**," "**Primary Ecological Succession**," and "**Succession: From Sand Dune to Ecology**" -- three videos representing various facets of succession.

JLM Visuals, 1208 Bridge Street, Grafton, WI 53204. --- "**Plant Succession**," "**Ecological Succession I**," "**Ecological Succession II**," and **Secondary Ecological Succession** - four titles for 35 mm. slide sets.